Sparse
Polynomial
Optimization

Theory and Practice

Series on Optimization and its Applications

Print ISSN: 2399-1593
Online ISSN: 2399-1607

(*formerly known as Imperial College Press Optimization Series* —
Print ISSN: 2041-1677)

Series Editor: Jean Bernard Lasserre *(LAAS-CNRS and Institute of
Mathematics, University of Toulouse, France)*

Published

Series on Optimization and its Applications – Vol. 5

Sparse
Polynomial
Optimization

Theory and Practice

Victor Magron
LAAS-CNRS, France

Jie Wang
Chinese Academy of Sciences, China

World Scientific

NEW JERSEY · LONDON · SINGAPORE · BEIJING · SHANGHAI · HONG KONG · TAIPEI · CHENNAI · TOKYO

Published by

World Scientific Publishing Europe Ltd.

57 Shelton Street, Covent Garden, London WC2H 9HE

Head office: 5 Toh Tuck Link, Singapore 596224

USA office: 27 Warren Street, Suite 401-402, Hackensack, NJ 07601

Library of Congress Cataloging-in-Publication Data

Names: Magron, Victor, author. | Wang, Jie (Mathematician), author.
Title: Sparse polynomial optimization : theory and practice / Victor Magron, Jie Wang.
Description: New Jersey : World Scientific Publishing Europe Ltd., 2023. |
    Series: Series on optimization and its applications, 2399-1593 ; Vol. 5 |
    Includes bibliographical references.
Identifiers: LCCN 2023002582 | ISBN 9781800612945 (hardcover) |
    ISBN 9781800612952 (ebook)
Subjects: LCSH: Mathematical optimization. | Polynomials. | Sparse matrices
Classification: LCC QA402.5 .M276 2023 | DDC 519.7--dc23/eng20230410
LC record available at https://lccn.loc.gov/2023002582

British Library Cataloguing-in-Publication Data
A catalogue record for this book is available from the British Library.

For any available supplementary material, please visit
https://www.worldscientific.com/worldscibooks/10.1142/Q0382#t=suppl

Contents

III   Term sparsity   97

List of acronyms

List of symbols

$\mathbb{N}$	$\{0,1,2,\dots\}$
$\mathbb{N}^*$	$\{1,2,\dots\}$
$\mathbb{Q}$	the field of rational numbers
$\mathbb{R}$	the field of real numbers
$\mathbb{C}$	the field of complex numbers
$\mathbf{0}$	the zero vector
$\mathbf{x} = (x_1,\dots,x_n)$	a tuple of real variables
$\mathrm{supp}(f)$	the support of the polynomial f
$[m]$	$\{1,2,\dots,m\}$
$|\cdot|$	the cardinality of a set or 1-norm of a vector
$\mathbb{R}^{n\times m}$	the set of $n \times m$ real matrices
$\mathbb{S}_n$	the set of real symmetric $n \times n$ matrices
$\mathbb{S}_n^+$	the set of $n \times n$ PSD matrices
$\langle \mathbf{A}, \mathbf{B} \rangle$	the trace of $\mathbf{AB}$ for $\mathbf{A}, \mathbf{B} \in \mathbb{S}_n$
$\mathbf{I}_n$	the $n \times n$ identity matrix
$\mathbf{M} \succeq 0$	$\mathbf{M}$ is a PSD matrix
F, G, H	graphs
$G(V, E)$	a graph with nodes V and edges E
$V(G)$ (resp. $E(G)$)	the node (resp. edge) set of the graph G
$\mathbf{B}_G$	the adjacency matrix of the graph G with unit diagonal
$G \subseteq H$	G is a subgraph of H
G'	a chordal extension of the graph G
$\mathbb{S}(G)$	the set of real symmetric matrices with sparsity pattern G
Π_G	the projection from $\mathbb{S}_{|V(G)|}$ to the subspace $\mathbb{S}(G)$
$\mathfrak{g} = \{g_1,\dots,g_m\}$	a set of polynomials defining the constraints
$\mathbb{R}[\mathbf{x}]$	the ring of real n-variate polynomials
$\mathbb{R}[\mathbf{x}]_{2d}$	the set of real n-variate polynomials of degree at most $2d$
$\Sigma[\mathbf{x}]$	the set of SOS polynomials
$\Sigma[\mathbf{x}]_d$	the set of SOS polynomials of degree at most $2d$

$\mathcal{M}(\mathfrak{g})$	the quadratic module generated by $\mathfrak{g}$
$\mathcal{M}(\mathfrak{g})_r$	the r-truncated quadratic module generated by $\mathfrak{g}$
$\mathbf{X}$	a basic semialgebraic set
μ, ν	measures
$\mathbb{N}_r^n$	$\{\alpha \in \mathbb{N}^n \mid \sum_{j=1}^n \alpha_j \leq r\}$
d_j	the ceil of half degree of $g_j \in \mathfrak{g}$
r	relaxation order
$r_{\min}$	minimum relaxation order
$\mathbf{y}$	a moment sequence
$L_{\mathbf{y}}$	the linear functional associated to $\mathbf{y}$
$\mathbf{M}_r(\mathbf{y})$	the r-th order moment matrix associated to $\mathbf{y}$
$\mathbf{M}_r(g\mathbf{y})$	the r-th order localizing matrix associated to $\mathbf{y}$ and g
$\delta_{\mathbf{a}}$	the Dirac measure centered at $\mathbf{a}$
p	number of variable cliques
s	sparse order
$\underline{x} = (x_1, \dots, x_n)$	a tuple of noncommutating variables
$\mathbb{R}\langle \underline{x} \rangle$	the ring of real nc n-variate polynomials
$\mathbf{W}_r$	the vector of nc monomials of degree at most r
$\Sigma\langle \underline{x} \rangle$	the set of SOHS polynomials
$\mathcal{D}_{\mathfrak{g}}$	the nc semialgebraic set associated to $\mathfrak{g}$

Preface

Consider the following list of problems arising from various distinct fields:

- Design certifiable algorithms for robust geometric perception in the presence of a large amount of outliers;

- Minimizing a sum of rational fractions to estimate the fundamental matrix in epipolar geometry;

- Computing the maximal roundoff error bound for the output of a numerical program;

- Certifying the robustness of a deep neural network;

- Computing the maximum violation level of Bell inequalities;

- Verifying the stability of a networked system or a control system under deadline constraints;

- Approximate stability regions of differential systems, such as reachable sets or positively invariant sets;

- Finding a maximum cut in a graph;

- Minimizing the generator fuel cost under alternative current power-flow constraints.

All these important applications related to computer vision, computer arithmetic, deep learning, entanglement in quantum information, graph theory and energy networks, can be successfully tackled within the framework of polynomial optimization, an emerging field with growing research efforts in the last two decades. One key advantage of these techniques is their ability to model a wide range of problems using optimization formulations. Polynomial optimization heavily relies on the moment-sums of squares (moment-SOS) approach proposed by Lasserre [Las01], which provides certificates for positive polynomials. The problem of minimizing a polynomial over a set of polynomial (in)-equalities is an NP-hard non-convex problem. It turns out that this problem can be cast as an infinite-dimensional linear problem over a set of probability measures. Thanks to powerful results from real algebraic geometry [Put93], one can convert this linear problem into a nested sequence of finite-dimensional convex problems. At each step of the associated hierarchy, one needs to solve a fixed size semidefinite program (an optimization program with a linear cost and constraints over matrices with nonnegative eigenvalues), which can be in turn solved with efficient numerical tools. On the practical side however, there is *no-free lunch* and such optimization methods usually encompass severe scalability issues. The underlying reason is that for optimization problems involving polynomials in n variables of degree at most $2d$, the size of the matrices involved at step $r \geq d$ of Lasserre's hierarchy of semidefinite programming (SDP) relaxations is proportional to $\binom{n+r}{r}$. Fortunately, for many applications, including the ones formerly mentioned, we can *look at the problem in the eyes* and exploit the inherent data structure arising from the cost and constraints describing the problem, for instance sparsity or symmetries.

> This book presents several research efforts to tackle this scientific challenge with important computational implications, and provides the development of alternative optimization schemes that scale well in terms of computational complexity, at least in some identified class of problems.

The presented algorithmic framework in this book mainly exploits the sparsity structure of the input data to solve large-scale polynomial optimization problems. For unconstrained problems involving a few terms, a first remedy consists of reducing the size of the relaxations by discarding the terms which never appear in the support of the sum of squares (SOS) decompositions. This technique, based on a result by Reznick [Rez78], consists of computing the Newton polytope of the input polynomial (the convex hull of the support of this polynomial) and selecting only monomials with supports lying in half of this polytope.

We present sparsity-exploiting hierarchies of relaxations, for either unconstrained or constrained polynomial optimization problems. By contrast with the dense hierarchies, they provide faster approximation of the solution in practice but also come with the same theoretical convergence guarantees. Our framework is not restricted to *static* polynomial optimization, and we expose hierarchies of approximations for values of interest arising from the analysis of dynamical systems. We also present various extensions to problems involving noncommuting variables, e.g., matrices of arbitrary size or quantum physic operators.

At this point, we would like to emphasize the existence of alterna-tives to the positivity certificates based on sparse SOS decompositions. Instead of computing SOS decompositions with SDP, one can compute other positivity certificates based on linear programming (LP) for Bern-stein decompositions or Krivine-Stengle certificates, geometric/second-order cone programming for nonnegative circuits and scaled diagonally dominant SOS, relative entropy programming for arithmetic-geometric-exponentials. This book also presents an overview of these various alter-native decompositions.

A second point to emphasize is that the concept of sparsity is inherent to many scientific fields, and we outline some similarities and differences with the algorithmic framework presented in this book. In the context of machine learning, statistics, or signal processing, exploiting sparsity boils down to select variables or features, usually with ℓ_1-norm regularization [BT09]. It is commonly employed to make the model or the prediction more interpretable or less expensive to use. In other words, even if the underlying problem does not admit sparse solutions, one still hopes to be able to find the best sparse approximation. A similar situation occurs in the context of dynamical systems with sparse state constraints and dy-namics, where the set of trajectories is not necessarily sparse. In the context of algebraic geometry, people have considered sparse systems of polyno-mial equations, where *sparse* means that the set of terms appearing in each equation is fixed. Bernshtein's theorem [Ber75] is a key ingredient as it pro-vides an accurate bound for the expected number of complex roots, based on the mixed volume of the Newton polytopes of polynomials describing the system. We similarly exploit support information given by Newton polytopes for our term-sparsity based hierarchies, presented in Part III.

This book is organized as follows:

Part I

Chapter 1 recalls some preliminary background on semidefinite program-ming, sparse matrix theory.

Chapter 2 outlines the basic concepts of the moment-SOS hierarchy in polynomial optimization.

Part II

This part of the book focuses on the notion of "correlative sparsity", occurring when there are few correlations between the variables of the input problem. This research investigation was initially developed by [WKKM06] and [Las06].

Chapter 3 is concerned with this first sparse variant of the moment-SOS hierarchy, based on correlative sparsity.

Chapter 4 explains how to apply the sparse moment-SOS hierarchy to provide efficiently upper bounds on roundoff errors of floating-point nonlinear programs.

Chapter 5 focuses on robustness certification of deep neural networks, in particular via Lipschitz constant estimation.

Chapter 6 describes a very distinct application for optimization of polynomials in noncommuting variables. We outline promising research perspectives in quantum information theory.

Part III

This part of the book presents a complementary framework, where we show how to exploit a distinct notion of sparsity, called "term sparsity", occurring when there are a small number of terms involved in the input problem by comparison with the fully dense case.

Chapter 7 focuses on this second sparse variant of the moment-SOS hierarchy, based on term sparsity.

Chapter 8 explains how to combine correlative and term sparsity.

Chapter 9 extends this term sparsity framework to complex polynomial optimization and shows how the resulting scheme can handle optimal power flow problems with tens of thousands of variables and constraints.

Chapter 10 extends the framework of exploiting term sparsity to noncommutative polynomial optimization (namely, eigenvalue optimization).

Chapter 11 is concerned with the application of this term sparsity framework to analyze the stability of various control systems, either coming from the networked systems literature or systems under deadline constraints.

Chapter 12 presents alternative algorithms to improve the scalability of polynomial optimization methods. First, we present algorithms based on sums of nonnegative circuit polynomials, recently introduced classes of nonnegativity certificates for sparse polynomials, which are independent of well-known methods based on sums of squares. Then, we outline existing methods to speed up the computation of the semidefinite relaxations.

Part IV

At the end of the book, we describe how to use various solvers available either in MATLAB or Julia. These dedicated appendices aim at guiding practitioners to solve optimization problems involving sparse polynomials.

Appendix A explains how to implement moment-SOS relaxations with software packages GloptiPoly and Yalmip.

Appendix B focuses on our sparsity exploiting algorithms, implemented in the TSSOS library available at https://github.com/wangjie212/TSSOS.

- For the sake of conciseness and clarity of exposition, most proofs are postponed to ease the reading. When the proof is either short or simple, we sometimes include it right after its corresponding statement. Otherwise, we refer to this proof in the *Notes and sources* section at the end of the corresponding chapter.

- Some of the theorems are framed in the book, in order to emphasize their specific importance.

Bibliography

[Ber75]   David N Bernshtein. The number of roots of a system of
            equations. *Functional Analysis and its applications*, 9(3):183–
            185, 1975.

[BT09]   Amir Beck and Marc Teboulle. A fast iterative shrinkage-
            thresholding algorithm for linear inverse problems. *SIAM
            journal on imaging sciences*, 2(1):183–202, 2009.

[Las01]   Jean-Bernard Lasserre. Global Optimization with Polynomi-
            als and the Problem of Moments. *SIAM Journal on Optimiza-
            tion*, 11(3):796–817, 2001.

[Las06]   Jean B Lasserre. Convergent sdp-relaxations in polynomial
            optimization with sparsity. *SIAM Journal on Optimization*,
            17(3):822–843, 2006.

[Put93]   M. Putinar. Positive polynomials on compact semi-algebraic
            sets. *Indiana University Mathematics Journal*, 42(3):969–984,
            1993.

[Rez78]   B. Reznick. Extremal PSD forms with few terms. *Duke Math-
            ematical Journal*, 45(2):363–374, 1978.

[WKKM06] Hayato Waki, Sunyoung Kim, Masakazu Kojima, and
            Masakazu Muramatsu. Sums of squares and semidefinite
            program relaxations for polynomial optimization problems
            with structured sparsity. *SIAM Journal on Optimization*,
            17(1):218–242, 2006.

Part I

Preliminary background

Chapter 1

Semidefinite programming and sparse matrices

In this chapter and the next one, we describe the foundations on which several parts of our work lie. Semidefinite programming and sparse matrices are described in this chapter while Chapter 2 is dedicated to the moment-SOS hierarchy of SDP relaxations, now widely used to certify lower bounds of polynomial optimization problems.

1.1 SDP and interior-point methods

Even though SDP is not our main topic of interest, several encountered problems can be cast as such programs.

First, we introduce some useful notations. We consider the vector space S_n of real symmetric $n \times n$ matrices, which is equipped with the usual inner product $\langle \mathbf{A}, \mathbf{B} \rangle = \operatorname{tr}(\mathbf{AB})$ for $\mathbf{A}, \mathbf{B} \in S_n$. Let $\mathbf{I}_n$ be the $n \times n$ identity matrix. A matrix $\mathbf{M} \in S_n$ is called *positive semidefinite (PSD)* (resp. *positive definite*) if $\mathbf{x}^\mathsf{T}\mathbf{Mx} \geq 0$ (resp. > 0), for all $\mathbf{x} \in \mathbb{R}^n$. In this case, we write $\mathbf{M} \succeq 0$ and define a partial order by writing $\mathbf{A} \succeq \mathbf{B}$ (resp. $\mathbf{A} \succ B$) if and only if $\mathbf{A} - \mathbf{B}$ is positive semidefinite (resp. positive definite). The set of $n \times n$ PSD matrices is denoted by S_n^+.

In semidefinite programming, one minimizes a linear objective function subject to a linear matrix inequality (LMI). The variable of the problem is the vector $\mathbf{y} \in \mathbb{R}^m$ and the input data of the problem are the vector $\mathbf{c} \in \mathbb{R}^m$ and symmetric matrices $\mathbf{F}_0, \dots, \mathbf{F}_m \in S_n$. The primal semidefinite program is defined as follows:

$$
\begin{aligned}
p_{\mathrm{sdp}} := \inf_{\mathbf{y} \in \mathbb{R}^m} \quad & \mathbf{c}^\mathsf{T}\mathbf{y} \\
\text{s.t.} \quad & \mathbf{F}(\mathbf{y}) \succeq 0
\end{aligned}
\tag{1.1}
$$

where

$$\mathbf{F}(\mathbf{y}) := \mathbf{F}_0 + \sum_{i=1}^{m} y_i \mathbf{F}_i.$$

The primal problem (1.1) is convex since the linear objective function and the linear matrix inequality constraint are both convex. We say that $\mathbf{y}$ is primal feasible (resp. strictly feasible) if $\mathbf{F}(\mathbf{y}) \succeq 0$ (resp. $\mathbf{F}(\mathbf{y}) \succ 0$). Furthermore, we associate the following dual problem with the primal problem (1.1):

$$\begin{aligned} d_{\mathrm{sdp}} := \sup_{\mathbf{G} \in \mathbb{S}_n} \quad & -\langle \mathbf{F}_0, \mathbf{G} \rangle \\ \text{s.t.} \quad & \langle \mathbf{F}_i, \mathbf{G} \rangle = c_i, \quad i \in [m] \\ & \mathbf{G} \succeq 0. \end{aligned} \tag{1.2}$$

The variable of the dual program (1.2) is the real symmetric matrix $\mathbf{G} \in \mathbb{S}_n$. We say that $\mathbf{G}$ is dual feasible (resp. strictly feasible) if $\langle \mathbf{F}_i, \mathbf{G} \rangle = c_i$, $i \in [m]$ and $\mathbf{G} \succeq 0$ (resp. $\mathbf{G} \succ 0$).

We will describe briefly the primal-dual interior-point method (used for instance by SDPA [YFN$^+$10], Mosek [ART03]), that solves the following primal-dual optimization problem:

$$\begin{cases} \inf_{\mathbf{y} \in \mathbb{R}^m, \mathbf{G} \in \mathbb{S}_n} \quad & \eta(\mathbf{y}, \mathbf{G}) \\ \text{s.t.} \quad & \langle \mathbf{F}_i, \mathbf{G} \rangle = c_i, \quad i \in [m] \\ & \mathbf{F}(\mathbf{y}) \succeq 0, \mathbf{G} \succeq 0 \end{cases} \tag{1.3}$$

where $\eta(\mathbf{y}, \mathbf{G}) := \mathbf{c}^\mathsf{T} \mathbf{y} + \langle \mathbf{F}_0, \mathbf{G} \rangle$.

We notice that the objective function η of the program (1.3) is the difference between the objective function of the primal program (1.1) and its dual version (1.2). We call this function the duality gap. Let us suppose that $\mathbf{y}$ is primal feasible and $\mathbf{G}$ is dual feasible, then η is nonnegative. Indeed, we have

$$\eta(\mathbf{y}, \mathbf{G}) = \sum_{i=1}^{m} \langle \mathbf{F}_i, \mathbf{G} \rangle y_i + \langle \mathbf{F}_0, \mathbf{G} \rangle = \langle \mathbf{F}(\mathbf{y}), \mathbf{G} \rangle \geq 0. \tag{1.4}$$

The last inequality comes from the fact that the matrices $\mathbf{F}(\mathbf{y})$ and $\mathbf{G}$ are both PSD.

Then, one can easily prove that the nonnegativity of η implies the following inequalities:

$$d_{\mathrm{sdp}} \leq -\langle \mathbf{F}_0, \mathbf{G} \rangle \leq \mathbf{c}^\mathsf{T} \mathbf{y} \leq p_{\mathrm{sdp}}. \tag{1.5}$$

Our problems that can be cast as SDPs satisfy certain assumptions, so that there exists a (strictly feasible) primal-dual optimal solution (i.e., a primal strictly feasible $\mathbf{y}$ solving (1.1) and a dual strictly feasible $\mathbf{G}$ solving (1.2)). Then, all inequalities in (1.5) become equalities and there is no duality gap ($\eta(\mathbf{y}, \mathbf{G}) = 0$):

$$d_{\text{sdp}} = -\langle \mathbf{F}_0, \mathbf{G} \rangle = \mathbf{c}^\mathsf{T}\mathbf{y} = p_{\text{sdp}}. \tag{1.6}$$

Thus, we will assume that such a primal-dual optimal solution exists in the sequel. We also introduce the barrier function

$$\Phi(\mathbf{y}) := \begin{cases} \log \det(\mathbf{F}(\mathbf{y})^{-1}) & \text{if } \mathbf{F}(\mathbf{y}) \succ 0, \\ +\infty & \text{otherwise.} \end{cases} \tag{1.7}$$

This barrier function exhibits several nice properties: Φ is strictly convex, analytic and self-concordant. The unique minimizer $\mathbf{y}^{\text{opt}}$ of Φ is called the analytic center of the LMI $\mathbf{F}(\mathbf{y}) \succeq 0$. This self-concordant barrier function guarantees that the number of iterations of the interior-point method is bounded by a polynomial in the dimension (n and m) and the number of accuracy digits of the solution.

1.2 Chordal graphs and sparse matrices

We briefly recall some basic notions from graph theory. An *(undirected) graph* $G(V, E)$ or simply G consists of a set of nodes V and a set of edges $E \subseteq \{\{v_i, v_j\} \mid v_i \neq v_j, (v_i, v_j) \in V \times V\}$. For a graph G, we use $V(G)$ and $E(G)$ to indicate the node set of G and the edge set of G, respectively. The *adjacency matrix* of a graph G is denoted by $\mathbf{B}_G$ for which we put ones on its diagonal. For two graphs G, H, we say that G is a *subgraph* of H if $V(G) \subseteq V(H)$ and $E(G) \subseteq E(H)$, denoted by $G \subseteq H$. For a graph $G(V, E)$, a *cycle* of length k is a set of nodes $\{v_1, v_2, \ldots, v_k\} \subseteq V$ with $\{v_k, v_1\} \in E$ and $\{v_i, v_{i+1}\} \in E$, for $i \in [k-1]$. A *chord* in a cycle $\{v_1, v_2, \ldots, v_k\}$ is an edge $\{v_i, v_j\}$ that joins two nonconsecutive nodes in the cycle. A *clique* $C \subseteq V$ of G is a subset of nodes where $\{v_i, v_j\} \in E$ for any $v_i, v_j \in C$. If a clique is not a subset of any other clique, then it is called a *maximal clique*.

Definition 1.1 (chordal graph) *A graph is called a* chordal graph *if all its cycles of length at least four have a chord.*

The notion of chordal graphs plays an important role in sparse matrix theory. In particular, it is known that maximal cliques of a chordal graph can be enumerated efficiently in linear time in the number of nodes and edges of the graph. See, e.g., [Gav72, VA+15a] for the details.

The maximal cliques $I_1, \ldots, I_p$ of a chordal graph (possibly after some reordering) satisfy the so-called *running intersection property (RIP)*, i.e., for every $k \in [p-1]$, it holds

$$\left(I_{k+1} \cap \bigcup_{j \leq k} I_j \right) \subseteq I_i \quad \text{for some } i \leq k. \tag{1.8}$$

The RIP actually gives an equivalent characterization of chordal graphs.

Theorem 1.2 *A connected graph is chordal if and only if its maximal cliques after an appropriate ordering satisfy the RIP.*

Any non-chordal graph $G(V, E)$ can always be extended to a chordal graph $G'(V, E')$ by adding appropriate edges to E, which is called a *chordal extension* of $G(V, E)$. The chordal extension of G is usually not unique. We use the symbol G' to indicate a specific chordal extension of G. For graphs $G \subseteq H$, we assume that $G' \subseteq H'$ always holds for our purpose. For a graph G, among all chordal extensions of G, there is a particular one G' which makes every connected component of G to be a clique. Accordingly, a matrix with adjacency graph G' is block diagonal (after an appropriate permutation on rows and columns): each block corresponds to a connected component of G. We call this chordal extension the *maximal* chordal extension. Besides, we are also interested in smallest chordal extensions. By definition, a *smallest chordal extension* is a chordal extension with the smallest clique number (i.e., the maximal size of maximal cliques). However, computing a smallest chordal extension is generally NP-complete [ACP87]. Therefore in practice we compute approximately smallest chordal extensions instead with efficient heuristic algorithms; see [BK10] for more detailed discussions.

Example 1.3 *Let us consider the graph $G(V, E)$ represented in Figure 1.1, with the set of nodes $V = \{1, 2, 3, 4, 5, 6\}$ and*

$$E = \{\{1,2\}, \{1,3\}, \{1,4\}, \{1,5\}, \{1,6\}, \{2,3\}, \{2,5\}, \{3,6\}, \{5,6\}\}$$

and the corresponding adjacency matrix

$$\mathbf{B}_G = \begin{bmatrix} 1 & 1 & 1 & 1 & 1 & 1 \\ 1 & 1 & 1 & 0 & 1 & 0 \\ 1 & 1 & 1 & 0 & 1 & 1 \\ 1 & 0 & 0 & 1 & 0 & 0 \\ 1 & 1 & 1 & 0 & 1 & 1 \\ 1 & 0 & 1 & 0 & 1 & 1 \end{bmatrix}.$$

One example of cycle of length 3 is $\{1, 5, 6\}$ and one example of cycle of length 4 is $\{6, 3, 2, 5\}$. Note that this graph is not chordal since there is no chord in this latter cycle. It is enough to add en edge between the nodes 2 and 6 (or alternatively between the nodes 3 and 5) to obtain a chordal extension of G.

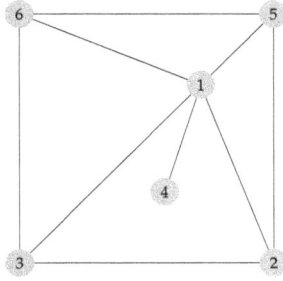

Figure 1.1: The graph from Example 1.3.

Let $n \in \mathbb{N}^*$. Given a graph $G(V, E)$ with $V = [n]$, a symmetric matrix $\mathbf{Q}$ with rows and columns indexed by V is said to have sparsity pattern G if $\mathbf{Q}_{ij} = \mathbf{Q}_{ij} = 0$ whenever $i \neq j$ and $\{i, j\} \notin E$. Let $S(G)$ be the set of real symmetric matrices with sparsity pattern G. The PSD matrices with sparsity pattern G form a convex cone

$$S^+_{|V|} \cap S(G) = \{\mathbf{Q} \in S(G) \mid \mathbf{Q} \succeq 0\}. \tag{1.9}$$

A matrix in $S(G)$ exhibits a block structure: each block corresponds to a maximal clique of G. Figure 1.2 depicts an instance of such block structures. Note that there might be overlaps between blocks because different maximal cliques may share nodes.

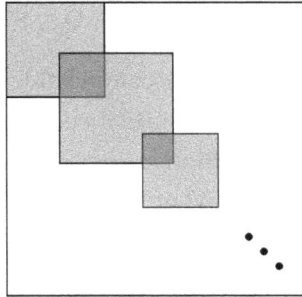

Figure 1.2: An instance of block structures for matrices in $S(G)$. The blue area indicates the positions of possible nonzero entries.

Given a maximal clique C of $G(V, E)$, we define a matrix $\mathbf{R}_C \in \mathbb{R}^{|C| \times |V|}$ by

$$[\mathbf{R}_C]_{ij} = \begin{cases} 1, & \text{if } C(i) = j, \\ 0, & \text{otherwise,} \end{cases} \tag{1.10}$$

where $C(i)$ denotes the i-th node in C, sorted with respect to the ordering compatible with V. Note that $\mathbf{Q}_C = \mathbf{R}_C \mathbf{Q} \mathbf{R}_C^\mathsf{T} \in S_{|C|}$ extracts a principal submatrix $\mathbf{Q}_C$ indexed by the clique C from a symmetry matrix $\mathbf{Q}$, and $\mathbf{Q} = \mathbf{R}_C^\mathsf{T} \mathbf{Q}_C \mathbf{R}_C$ inflates a $|C| \times |C|$ matrix $\mathbf{Q}_C$ into a sparse $|V| \times |V|$ matrix $\mathbf{Q}$.

When the sparsity pattern graph G is chordal, the cone $\mathbb{S}_{|V|}^+ \cap \mathbb{S}(G)$ can be decomposed as a sum of simple convex cones, as stated in the following theorem.

> **Theorem 1.4** *Let $G(V, E)$ be a chordal graph and assume that $C_1, \ldots, C_p$ are the list of maximal cliques of G. Then a matrix $\mathbf{Q} \in \mathbb{S}_{|V|}^+ \cap \mathbb{S}(G)$ if and only if there exist $\mathbf{Q}_k \in \mathbb{S}_{|C_k|}^+$ for $k \in [p]$ such that $\mathbf{Q} = \sum_{k=1}^{p} \mathbf{R}_{C_k}^\mathsf{T} \mathbf{Q}_k \mathbf{R}_{C_k}$.*

Given a graph $G(V, E)$ with $V = [n]$, let Π_G be the projection from $\mathbb{S}_{|V|}$ to the subspace $\mathbb{S}(G)$, i.e., for $\mathbf{Q} \in \mathbb{S}_{|V|}$,

$$[\Pi_G(\mathbf{Q})]_{ij} = \begin{cases} \mathbf{Q}_{ij}, & \text{if } i = j \text{ or } \{i, j\} \in E, \\ 0, & \text{otherwise.} \end{cases} \tag{1.11}$$

We denote by $\Pi_G(\mathbb{S}_{|V|}^+)$ the set of matrices in $\mathbb{S}(G)$ that have a PSD completion, i.e.,

$$\Pi_G(\mathbb{S}_{|V|}^+) = \left\{ \Pi_G(\mathbf{Q}) \mid \mathbf{Q} \in \mathbb{S}_{|V|}^+ \right\}. \tag{1.12}$$

One can easily check that the PSD completable cone $\Pi_G(\mathbb{S}_{|V|}^+)$ and the PSD cone $\mathbb{S}_{|V|}^+ \cap \mathbb{S}(G)$ form a pair of dual cones in $\mathbb{S}(G)$. Moreover, for a chordal graph G, the decomposition result for the cone $\mathbb{S}_{|V|}^+ \cap \mathbb{S}(G)$ in Theorem 1.4 leads to the following characterization of the PSD completable cone $\Pi_G(\mathbb{S}_{|V|}^+)$.

> **Theorem 1.5** *Let $G(V, E)$ be a chordal graph and assume that $C_1, \ldots, C_p$ are the list of maximal cliques of G. Then a matrix $\mathbf{Q} \in \Pi_G(\mathbb{S}_{|V|}^+)$ if and only if $\mathbf{Q}_k = \mathbf{R}_{C_k} \mathbf{Q} \mathbf{R}_{C_k}^\mathsf{T} \succeq 0$ for all $k \in [p]$.*

Theorem 1.4 and Theorem 1.5 play an important role in sparse semidefinite programming since they admit us to decompose an SDP with chordal sparsity pattern into an SDP of smaller size, which yields significant computational improvement if the sizes of related maximal cliques are small.

1.3   Notes and sources

SDP is relevant to a wide range of applications. The interested reader can find more details on the connection between SDP and combinatorial optimization in [GLV09], control theory in [BEGFB94], positive semidefinite matrix completion in [Lau09]. A survey on semidefinite programming is available in the paper of Vandenberghe and Boyd [VB96]. We emphasize the fact that SDPs can be solved efficiently by software e.g., SeDuMi [Stu99], CSDP [Bor99], SDPA [YFN$^+$10], Mosek [ART03].

We refer to [NN94] for more details on barrier functions. Detailed complexity bounds related to SDP solving with interior-point methods can be found in Section 4.6.3 from [BTN01]. With prescribed accuracy, the time complexity of SDP (in terms of arithmetic operations) is polynomial with respect to the number of variables with an exponent greater than 3; see [BTN01, Chapter 4] for more details.

For more details about sparse matrices and chordal graphs, the reader is referred to the survey [VA$^+$15b]. Theorem 1.4 and Theorem 1.5 are stated as Theorem 9.2 and Theorem 10.1 in [VA$^+$15b], respectively, and were derived much earlier in [AHMR88] and [GJSW84], respectively.

The equivalence stated in Theorem 1.2 could be read from Theorem 3.4 or Corollary 1 of [BP93].

Bibliography

[ACP87]   Stefan Arnborg, Derek G Corneil, and Andrzej Proskurowski. Complexity of finding embeddings in a k-tree. *SIAM Journal on Algebraic Discrete Methods*, 8(2):277–284, 1987.

[AHMR88] Jim Agler, William Helton, Scott McCullough, and Leiba Rodman. Positive semidefinite matrices with a given sparsity pattern. *Linear algebra and its applications*, 107:101–149, 1988.

[ART03]   Erling D Andersen, Cornelis Roos, and Tamas Terlaky. On implementing a primal-dual interior-point method for conic quadratic optimization. *Mathematical Programming*, 95(2):249–277, 2003.

[BEGFB94] Stephen Boyd, Laurent El Ghaoui, Eric Feron, and Venkataramanan Balakrishnan. *Linear matrix inequalities in system and control theory*. SIAM, 1994.

[BK10]   Hans L Bodlaender and Arie MCA Koster. Treewidth computations i. upper bounds. *Information and Computation*, 208(3):259–275, 2010.

[Bor99]   Brian Borchers. Csdp, ac library for semidefinite programming. *Optimization methods and Software*, 11(1-4):613–623, 1999.

[BP93]   Jean RS Blair and Barry Peyton. An introduction to chordal graphs and clique trees. In *Graph theory and sparse matrix computation*, pages 1–29. Springer, 1993.

[BTN01]   Aharon Ben-Tal and Arkadi Nemirovski. *Lectures on modern convex optimization: analysis, algorithms, and engineering applications*. SIAM, 2001.

[Gav72]   Fănică Gavril. Algorithms for minimum coloring, maximum clique, minimum covering by cliques, and maximum independent set of a chordal graph. *SIAM Journal on Computing*, 1(2):180–187, 1972.

[GJSW84]   Robert Grone, Charles R Johnson, Eduardo M Sá, and Henry
           Wolkowicz. Positive definite completions of partial hermitian
           matrices. *Linear algebra and its applications*, 58:109–124, 1984.

[GLV09]   Nebojša Gvozdenović, Monique Laurent, and Frank Vallentin.
           Block-diagonal semidefinite programming hierarchies for 0/1
           programming. *Operations Research Letters*, 37(1):27–31, 2009.

[Lau09]   Monique Laurent. Matrix completion problems. *Encyclopedia
           of Optimization*, 3:221–229, 2009.

[NN94]   Yurii Nesterov and Arkadii Nemirovskii. *Interior-point polyno-
           mial algorithms in convex programming*. SIAM, 1994.

[Stu99]   Jos F Sturm. Using sedumi 1.02, a matlab toolbox for optimiza-
           tion over symmetric cones. *Optimization methods and software*,
           11(1-4):625–653, 1999.

[VA$^+$15a] Lieven Vandenberghe, Martin S Andersen, et al.   Chordal
           graphs and semidefinite optimization.   *Foundations and
           Trends® in Optimization*, 1(4):241–433, 2015.

[VA$^+$15b] Lieven Vandenberghe, Martin S Andersen, et al.   Chordal
           graphs and semidefinite optimization.   *Foundations and
           Trends® in Optimization*, 1(4):241–433, 2015.

[VB96]   Lieven Vandenberghe and Stephen Boyd. Semidefinite pro-
           gramming. *SIAM review*, 38(1):49–95, 1996.

[YFN$^+$10] M. Yamashita, K. Fujisawa, K. Nakata, M. Nakata, M. Fukuda,
           K. Kobayashi, and K. Goto.   A high-performance software
           package for semidefinite programs : SDPA7. Technical report,
           Dept. of Information Sciences, Tokyo Inst. Tech., 2010.

Chapter 2

Polynomial optimization and the moment-SOS hierarchy

Polynomial optimization focuses on minimizing or maximizing a polynomial under a set of polynomial inequality constraints. A polynomial is an expression involving addition, subtraction and multiplication of variables and coefficients. An example of polynomial in two variables x_1 and x_2 with rational coefficients is $f(x_1, x_2) = 1/3 + x_1^2 + 2x_1x_2 + x_2^2$. *Semialgebraic* sets are defined with conjunctions and disjunctions of polynomial inequalities with real coefficients. For instance the two-dimensional unit disk is a semialgebraic set defined as the set of all points (x_1, x_2) satisfying the (single) inequality $1 - x_1^2 - x_2^2 \geq 0$.

In general, computing the *exact* solution of a polynomial optimization problem (POP) over a semialgebraic set is an NP-hard problem. In practice, one can at least try to compute an *approximation* of the solution by considering a *relaxation* of the problem instead of the problem itself. The approximated solution may not satisfy all the problem constraints but still gives useful information about the exact solution. Let us illustrate this by considering the minimization of the above polynomial $f(x_1, x_2)$ on the unit disk. One can replace this disk by a larger set, for instance the product of intervals $[-1, 1] \times [-1, 1]$. Using basic interval arithmetic, one easily shows that the range of f belongs to $[-4/3, 4/3]$. Next, one can replace the monomials x_1^2, x_1x_2 and x_2^2 by three new variables y_1, y_2 and y_3, respectively. One can relax the initial problem by LP, with a cost of $1/3 + y_1 + 2y_2 + y_3$ and one single linear inequality constraint $1 - y_1 - y_3 \geq 0$. By hand-solving or by using an LP solver, one finds again a lower bound of $-4/3$. Even if LP gives more accurate bounds than interval arithmetic in general, this does not yield any improvement on this example.

One way to obtain more accurate lower bounds is to rely on more sophisticated techniques from the field of convex optimization, e.g., SDP. In the seminal paper [Las01] published in 2001, Lasserre introduced a hierarchy of relaxations allowing to obtain a converging sequence of lower bounds for the minimum of a polynomial over a semialgebraic set. Each lower bound is computed by SDP.

The idea behind Lasserre's hierarchy is to tackle the *infinite-dimensional* initial problem by solving several *finite-dimensional* primal-dual SDP problems. The primal is a *moment* problem, that is an optimization problem where variables are the moments of a Borel measure. The first moment is related to means, the second moment is related to variances, etc. Lasserre showed in [Las01] that POP can be cast as a particular instance of the generalized moment problem (GMP). In a nutshell, the primal moment problem approximates Borel measures. The dual is an SOS problem, where the variables are the coefficients of SOS polynomials (e.g., $(1/\sqrt{3})^2 + (x_1 + x_2)^2$). It is known that not all positive polynomials (on a semialgebraic set) can be written with SOS decompositions. However, if the set of constraints satisfies certain assumptions (slightly stronger than compactness), then one can represent positive polynomials with weighted SOS decompositions. In a nutshell, the dual SOS problem approximates positive polynomials. The moment-SOS approach can be used on the above example with either three moment variables or SOS of degree 2 to obtain a lower bound of $1/3$. For this example, the exact solution is obtained at the first step of the hierarchy. There is no need to go further, i.e., to consider primal with moments of greater order (e.g. the integrals of x_1^3, $x_1^2 x_2$, x_1^4) or dual with SOS polynomials of degree 4 or 6. The reason is that for convex quadratic problems, the first step of the hierarchy gives the exact solution!

In the sequel, we recall more formally some preliminary background material on the mathematical tools related to the moment-SOS hierarchy.

Given $n, d \in \mathbb{N}$, let $\mathbb{R}[\mathbf{x}]$ (resp. $\mathbb{R}[\mathbf{x}]_{2d}$) stand for the vector space of real n-variate polynomials (resp. of degree at most $2d$) in the variable $\mathbf{x} = (x_1, \ldots, x_n) \in \mathbb{R}^n$. A polynomial $f \in \mathbb{R}[\mathbf{x}]$ can be written as $f(\mathbf{x}) = \sum_{\alpha \in \mathbf{A}} f_\alpha \mathbf{x}^\alpha$ with $\mathbf{A} \subseteq \mathbb{N}^n$ and $f_\alpha \in \mathbb{R}, \mathbf{x}^\alpha = x_1^{\alpha_1} \cdots x_n^{\alpha_n}$. The *support* of f is defined by $\mathrm{supp}(f) := \{\alpha \in \mathbf{A} \mid f_\alpha \neq 0\}$. A basic compact semialgebraic set $\mathbf{X}$ is a finite conjunction of polynomial superlevel sets. Namely, given $m \in \mathbb{N}^*$ and a set of polynomials $\mathfrak{g} := \{g_1, \ldots, g_m\} \subset \mathbb{R}[\mathbf{x}]$, one has

$$
\begin{aligned}
\mathbf{X} &:= \{\mathbf{x} \in \mathbb{R}^n : g(\mathbf{x}) \geq 0 \text{ for all } g \in \mathfrak{g}\} \\
&= \{\mathbf{x} \in \mathbb{R}^n : g_1(\mathbf{x}) \geq 0, \ldots, g_m(\mathbf{x}) \geq 0\}.
\end{aligned} \tag{2.1}
$$

Many sets can be described as such basic closed semialgebraic sets, and the related description is not unique. Consider for instance the 2-dimensional hypercube $\mathbf{X} = [0,1]^2$. A first possible description is given by

$$\mathbf{X} = \{(x_1, x_2) \in \mathbb{R}^2 : x_1 - x_1^2 \geq 0, x_2 - x_2^2 \geq 0\} \tag{2.2}$$
$$= \{(x_1, x_2) \in \mathbb{R}^2 : g(x_1, x_2) \geq 0 \text{ for all } g \in \mathfrak{g}\},$$

with $\mathfrak{g} = \{x_1 - x_1^2, x_2 - x_2^2\}$. A second one is given by taking $\mathfrak{g} = \{x_1, 1 - x_1, x_2, 1 - x_2\}$.

2.1   Sums of squares and quadratic modules

Let $\Sigma[\mathbf{x}]$ stand for the cone of SOS polynomials and let $\Sigma[\mathbf{x}]_d$ denote the cone of SOS polynomials of degree at most $2d$, namely $\Sigma[\mathbf{x}]_d := \Sigma[\mathbf{x}] \cap \mathbb{R}[\mathbf{x}]_{2d}$. Note that all SOS polynomials with real coefficients are nonnegative on $\mathbb{R}^n$. For instance, $\sigma_0 = \frac{1}{2}(x_1 + x_2 - \frac{1}{2})^2$ is a square in $n = 2$ variables of degree $2d = 2$, and is thus obviously nonnegative on $\mathbb{R}^n$.

For the ease of further notation, we set $g_0 := 1$, and $d_j := \lceil \deg(g_j)/2 \rceil$, for all $j = 0, \ldots, m$. Given a basic compact semialgebraic set $\mathbf{X}$ as above and an integer $r \in \mathbb{N}^*$, let $\mathcal{M}(\mathfrak{g})$ be the quadratic module generated by $g_1, \ldots, g_m$:

$$\mathcal{M}(\mathfrak{g}) := \left\{ \sum_{j=0}^{m} \sigma_j(\mathbf{x}) g_j(\mathbf{x}) : \sigma_j \in \Sigma[\mathbf{x}], j = 0, \ldots, m \right\},$$

and let $\mathcal{M}(\mathfrak{g})_r$ be the r-truncated quadratic module:

$$\mathcal{M}(\mathfrak{g})_r := \left\{ \sum_{j=0}^{m} \sigma_j(\mathbf{x}) g_j(\mathbf{x}) : \sigma_j \in \Sigma[\mathbf{x}]_{r-d_j}, j = 0, \ldots, m \right\}.$$

A first important remark is that all polynomials belonging to $\mathcal{M}(\mathfrak{g})$ are positive on $\mathbf{X}$.

Example 2.1 *To illustrate this later point, let us take the polynomial $f = x_1 x_2$ in two variables, and the 2-dimensional hypercube $\mathbf{X} = [0,1]^2$ described with the basic closed semialgebraic set given in (2.2), with $\mathfrak{g} = \{x_1(1 - x_1), x_2(1 - x_2)\}$. Let us consider the following decomposition of $f + \frac{1}{8}$:*

$$x_1 x_2 + \frac{1}{8} = \frac{1}{2}\left(x_1 + x_2 - \frac{1}{2}\right)^2 \cdot 1 + \frac{1}{2} \cdot x_1(1 - x_1) + \frac{1}{2} \cdot x_2(1 - x_2).$$

This later decomposition of degree 2 proves that $f = x_1 x_2 + \frac{1}{8}$ lies in the 1-truncated quadratic module $\mathcal{M}(\mathfrak{g})_1$, with $\sigma_0 = \frac{1}{2}(x_1 + x_2 - \frac{1}{2})^2$, $\sigma_1 = \sigma_2 = \frac{1}{2}$, $g_0 = 1$, $g_1 = x_1(1 - x_1)$ and $g_2 = x_2(1 - x_2)$. Since $\sigma_0, \sigma_1, \sigma_2$ are nonnegative on $\mathbb{R}^2$ and g_1, g_2 are nonnegative on $\mathbf{X}$ (by definition), the above decomposition certifies *that $f + \frac{1}{8} \geq 0$ on the hypercube $[0,1]^2$. This yields in particular that* $-\frac{1}{8}$ *is a lower bound on the minimum of f on the hypercube. Since the minimum of f on the hypercube is obviously 0, it is natural to ask if one can compute a lower bound greater than* $-\frac{1}{8}$. *The answer is positive: for all arbitrary small $\varepsilon > 0$, there exists a decomposition of $f + \varepsilon$ in $\mathcal{M}(\mathfrak{g})_r$, for some positive integer r depending on ε.*

A second important remark is that $\mathcal{M}(\mathfrak{g})_r \subseteq \mathcal{M}(\mathfrak{g})_{r+1}$, for all $r \in \mathbb{N}^*$, since all SOS polynomials of degree $2r$ can be viewed as SOS polynomials of degree $2r + 2$.

To guarantee the convergence behavior of the relaxations presented in the sequel, we rely on the fact that polynomials which are positive on $\mathbf{X}$ lie in $\mathcal{M}(\mathfrak{g})_r$ for some $r \in \mathbb{N}^*$. The existence of such SOS-based representations is guaranteed when the following condition holds.

Assumption 2.2 (Archimedean) *There exists $N > 0$ such that $N - \|\mathbf{x}\|_2^2 \in \mathcal{M}(\mathfrak{g})$.*

A quadratic module $\mathcal{M}(\mathfrak{g})$ for which Assumption 2.2 holds is said to be *Archimedean.*

Theorem 2.3 (Putinar's Positivstellensatz) *Assume that the set $\mathbf{X}$ is defined in (2.1) and the quadratic module $\mathcal{M}(\mathfrak{g})$ is Archimedean. Then any polynomial positive on $\mathbf{X}$ belongs to $\mathcal{M}(\mathfrak{g})$.*

Assumption 2.2 is slightly stronger than compactness. Indeed, compactness of $\mathbf{X}$ already ensures that each variable has finite lower and upper bounds. One (easy) way to ensure that Assumption 2.2 holds is to add a redundant constraint involving a well-chosen N depending on these bounds, in the definition of $\mathbf{X}$.

2.2   Borel measures and moment matrices

Given a compact set $\mathbf{A} \subseteq \mathbb{R}^n$, we denote by $\mathcal{M}(\mathbf{A})$ the vector space of finite signed Borel measures supported on $\mathbf{A}$, namely real-valued functions from the Borel σ-algebra $\mathcal{B}(\mathbf{A})$. The support of a measure $\mu \in \mathcal{M}(\mathbf{A})$ is defined as the closure of the set of all points $\mathbf{x}$ such that $\mu(\mathbf{B}) \neq 0$ for any open neighborhood $\mathbf{B}$ of $\mathbf{x}$. We denote by $\mathscr{C}(\mathbf{A})$ the Banach space of continuous functions on $\mathbf{A}$ equipped with the sup-norm. Let $\mathscr{C}(\mathbf{A})'$ stand for the topological dual of $\mathscr{C}(\mathbf{A})$ (equipped with the sup-norm), i.e., the set of continuous linear functionals of $\mathscr{C}(\mathbf{A})$. By a Riesz identification theorem, $\mathscr{C}(\mathbf{A})'$ is isomorphically identified with $\mathcal{M}(\mathbf{A})$ equipped with the total variation norm denoted by $\| \cdot \|_{\text{TV}}$. Let $\mathscr{C}_+(\mathbf{A})$ (resp. $\mathcal{M}_+(\mathbf{A})$) stand for the cone of nonnegative elements of $\mathscr{C}(\mathbf{A})$ (resp. $\mathcal{M}(\mathbf{A})$). The topology in $\mathscr{C}_+(\mathbf{A})$ is the strong topology of uniform convergence in contrast with the weak-star topology in $\mathcal{M}_+(\mathbf{A})$.

With $\mathbf{X}$ being a basic compact semialgebraic set, the restriction of the Lebesgue measure on a subset $\mathbf{A} \subseteq \mathbf{X}$ is $\lambda_{\mathbf{A}}(\mathbf{dx}) := \mathbf{1}_{\mathbf{A}}(\mathbf{x})\mathbf{dx}$, where $\mathbf{1}_{\mathbf{A}} : \mathbf{X} \to \{0,1\}$ stands for the indicator function of $\mathbf{A}$, namely $\mathbf{1}_{\mathbf{A}}(\mathbf{x}) = 1$ if $\mathbf{x} \in \mathbf{A}$ and $\mathbf{1}_{\mathbf{A}}(\mathbf{x}) = 0$ otherwise. A sequence $\mathbf{y} := (y_\alpha)_{\alpha \in \mathbb{N}^n} \subseteq \mathbb{R}$ is said to have a representing measure on $\mathbf{X}$ if there exists $\mu \in \mathscr{M}(\mathbf{X})$ such that $y_\alpha = \int \mathbf{x}^\alpha \mu(\mathbf{dx})$ for all $\alpha \in \mathbb{N}^n$, where we use the multinomial notation $\mathbf{x}^\alpha := x_1^{\alpha_1} x_2^{\alpha_2} \cdots x_n^{\alpha_n}$.

Assume that $\mu, \nu \in \mathscr{M}_+(\mathbf{X})$ have the same moments $\mathbf{y}$, namely $y_\alpha = \int_{\mathbf{X}} \mathbf{x}^\alpha \, d\mu = \int_{\mathbf{X}} \mathbf{x}^\alpha \, d\nu$, for all $\alpha \in \mathbb{N}^n$. Let us fix $f \in \mathscr{C}(\mathbf{X})$. Since $\mathbf{X}$ is compact, the Stone-Weierstrass theorem implies that the polynomials are dense in $\mathscr{C}(\mathbf{X})$, so $\int_{\mathbf{X}} f \, d\mu = \int_{\mathbf{X}} f \, d\nu$. Since f was arbitrary, the above equality holds for any $f \in \mathscr{C}(\mathbf{X})$, which implies that $\mu = \nu$. Therefore, any finite Borel measures supported on $\mathbf{X}$ is *moment determinate*.

The moments of the Lebesgue measure on $\mathbf{A}$ are denoted by

$$y_\alpha^{\mathbf{A}} := \int \mathbf{x}^\alpha \lambda_{\mathbf{A}} \, dx \in \mathbb{R}, \quad \alpha \in \mathbb{N}^n. \tag{2.3}$$

The Lebesgue volume of $\mathbf{A}$ is $\mathrm{vol}\, \mathbf{A} := y_0^{\mathbf{A}} = \int \lambda_{\mathbf{A}} \, dx$. For all $r \in \mathbb{N}$, let us define $\mathbb{N}_r^n := \{\alpha \in \mathbb{N}^n \mid \sum_{j=1}^n \alpha_j \leq r\}$, whose cardinality is $\binom{n+r}{r}$. Then a polynomial $f \in \mathbb{R}[\mathbf{x}]$ is written as follows:

$$\mathbf{x} \mapsto f(\mathbf{x}) = \sum_{\alpha \in \mathbb{N}^n} f_\alpha \mathbf{x}^\alpha,$$

and f is identified with its vector of coefficients $\mathbf{f} = (f_\alpha)_{\alpha \in \mathbb{N}^n}$ in the standard monomial basis $(\mathbf{x}^\alpha)_{\alpha \in \mathbb{N}^n}$.

Given a real sequence $\mathbf{y} = (y_\alpha)_{\alpha \in \mathbb{N}^n}$, let us define the linear functional $L_{\mathbf{y}} : \mathbb{R}[\mathbf{x}] \to \mathbb{R}$ by $L_{\mathbf{y}}(f) := \sum_\alpha f_\alpha y_\alpha$, for every polynomial $f = \sum_\alpha f_\alpha \mathbf{x}^\alpha$. Coming back to the previous 2-dimensional example from Section 2.1, with $f = x_1 x_2$, $g_1 = x_1 - x_1^2$ and $g_2 = x_2 - x_2^2$, we have $L_{\mathbf{y}}(f) = y_{11}$, $L_{\mathbf{y}}(g_1) = y_{10} - y_{20}$ and $L_{\mathbf{y}}(g_2) = y_{01} - y_{02}$.

Then, we associate to $\mathbf{y}$ the so-called *moment matrix* $\mathbf{M}_r(\mathbf{y})$ of order r, that is the real symmetric matrix with rows and columns indexed by $\mathbb{N}_r^n$ and the following entrywise definition:

$$[\mathbf{M}_r(\mathbf{y})]_{\beta,\gamma} := L_{\mathbf{y}}(\mathbf{x}^{\beta+\gamma}), \quad \forall \beta, \gamma \in \mathbb{N}_r^n.$$

Given $g \in \mathbb{R}[\mathbf{x}]$, we also associate to $\mathbf{y}$ and g the so-called *localizing matrix* of order r, that is the real symmetric matrix $\mathbf{M}_r(g\,\mathbf{y})$ with rows and columns indexed by $\mathbb{N}_r^n$ and the following entrywise definition:

$$[\mathbf{M}_r(g\,\mathbf{y})]_{\beta,\gamma} := L_{\mathbf{y}}(g(\mathbf{x})\,\mathbf{x}^{\beta+\gamma}), \quad \forall \beta, \gamma \in \mathbb{N}_r^n.$$

Let $\mathbf{X}$ be a basic compact semialgebraic set as in (2.1). Then one can check that if $\mathbf{y}$ has a representing measure $\mu \in \mathscr{M}_+(\mathbf{X})$ then $\mathbf{M}_r(g_j\,\mathbf{y}) \succeq 0$, for all $j = 0, \ldots, m$.

Let us give a simple example to illustrate the construction of moment and localizing matrices.

Example 2.4 *Let us take $n = 2$ and $r = 2$. The moment matrix $\mathbf{M}_2(\mathbf{y})$ is indexed by $\mathbb{N}_2^2 = \{(0,0), (1,0), (0,1), (2,0), (1,1), (0,2)\}$ and can be written as follows:*

$$\mathbf{M}_2(\mathbf{y}) = \begin{bmatrix} 1 & | & y_{1,0} & y_{0,1} & | & y_{2,0} & y_{1,1} & y_{0,2} \\ \hline y_{1,0} & | & y_{2,0} & y_{1,1} & | & y_{3,0} & y_{2,1} & y_{1,2} \\ y_{0,1} & | & y_{1,1} & y_{0,2} & | & y_{2,1} & y_{1,2} & y_{0,3} \\ \hline y_{2,0} & | & y_{3,0} & y_{2,1} & | & y_{4,0} & y_{3,1} & y_{2,2} \\ y_{1,1} & | & y_{2,1} & y_{1,2} & | & y_{3,1} & y_{2,2} & y_{1,3} \\ y_{0,2} & | & y_{1,2} & y_{0,3} & | & y_{2,2} & y_{1,3} & y_{0,4} \end{bmatrix}.$$

Next, consider the polynomial $g_1(\mathbf{x}) = x_1 - x_1^2$ of degree 2. From the first-order moment matrix:

$$\mathbf{M}_1(\mathbf{y}) = \begin{bmatrix} 1 & | & y_{1,0} & y_{0,1} \\ \hline y_{1,0} & | & y_{2,0} & y_{1,1} \\ y_{0,1} & | & y_{1,1} & y_{0,2} \end{bmatrix},$$

we obtain the following localizing matrix:

$$\mathbf{M}_1(g_1\mathbf{y}) = \begin{bmatrix} y_{1,0} - y_{2,0} & y_{2,0} - y_{3,0} & y_{1,1} - y_{2,1} \\ y_{2,0} - y_{3,0} & y_{3,0} - y_{4,0} & y_{2,1} - y_{3,1} \\ y_{1,1} - y_{2,1} & y_{2,1} - y_{3,1} & y_{1,2} - y_{2,2} \end{bmatrix}.$$

For instance, the last entry $[\mathbf{M}_1(g_1\mathbf{y})]_{3,3}$ is equal to

$$L_\mathbf{y}(g_1(\mathbf{x}) \cdot x_2 \cdot x_2) = L_\mathbf{y}(x_1 x_2^2 - x_1^2 x_2^2) = y_{1,2} - y_{2,2}.$$

2.3   The moment-SOS hierarchy

Let us consider the general POP

$$\mathbf{P}: \quad f_{\min} = \inf_\mathbf{x} \{f(\mathbf{x}) : \mathbf{x} \in \mathbf{X}\}, \tag{2.4}$$

where f is a polynomial and $\mathbf{X}$ is a basic closed semialgebraic set as in (2.1). It happens that this problem can be cast as an LP over probability measures, namely,

$$f_{\text{meas}} := \inf_{\mu \in \mathcal{M}_+(\mathbf{X})} \left\{ \int_\mathbf{X} f \, d\mu : \int_\mathbf{X} d\mu = 1 \right\}. \tag{2.5}$$

To see that $f_{meas} = f_{min}$ holds, let us consider a global minimizer $\mathbf{x}^{opt} \in \mathbf{R}^n$ of f on $\mathbf{X}$ and consider the Dirac measure $\mu = \delta_{\mathbf{x}^{opt}}$ supported on this point. Note that this Dirac (probability) measure is feasible for the LP stated in (2.5), with value $\int_{\mathbf{X}} f \, d\mu = f(\mathbf{x}^{opt}) = f_{min}$, which implies that $\inf_{\mu \in \mathcal{M}_+(\mathbf{X})} \{ \int_{\mathbf{X}} f \, d\mu : \int_{\mathbf{X}} d\mu = 1 \} \leq f_{min}$. For the other direction, let us consider a measure μ feasible for LP (2.5). Then, simply observe that since $f(\mathbf{x}) \geq f_{min}$, for all $\mathbf{x} \in \mathbf{X}$, the feasibility of μ implies that $\int_{\mathbf{X}} f \, d\mu \geq \int_{\mathbf{X}} f_{min} \, d\mu = f_{min} \int_{\mathbf{X}} d\mu = f_{min}$. Since it is true for any feasible solution, one has $\inf_{\mu \in \mathcal{M}_+(\mathbf{X})} \{ \int_{\mathbf{X}} f \, d\mu : \int_{\mathbf{X}} d\mu = 1 \} \geq f_{min}$. Another way to state this equality is to write

$$f_{min} = \sup_{b} \{ b : f - b \geq 0 \text{ on } \mathbf{X} \}, \tag{2.6}$$

which is an LP over nonnegative polynomials, and to notice that the dual LP of (2.6) is LP (2.5). The equality then follows from the zero duality gap in infinite-dimensional LP.

After reformulating **P** as LP (2.5) over probability measures, one can then build a hierarchy of moment relaxations for the later problem. This is done by using the fact that the condition $\mu \in \mathcal{M}_+(\mathbf{X})$ can be relaxed as $\mathbf{M}_{r-d_j}(g_j \, \mathbf{y}) \succeq 0$, for all $j = 0, \ldots, m$, and all $r \geq d_j = \lceil \deg(g_j)/2 \rceil$.

Letting $r_{min} := \max \{ \lceil \deg(f)/2 \rceil, d_1, \ldots, d_m \}$, at step $r \geq r_{min}$ of the hierarchy, one considers the following primal SDP program:

$$\mathbf{P}^r : \qquad \begin{aligned} f^r := \quad &\inf_{\mathbf{y}} \quad L_{\mathbf{y}}(f) \\ &\text{s.t.} \quad \mathbf{M}_r(\mathbf{y}) \succeq 0 \\ &\qquad \mathbf{M}_{r-d_j}(g_j \, \mathbf{y}) \succeq 0, \quad j \in [m] \\ &\qquad y_0 = 1. \end{aligned} \tag{2.7}$$

Before considering the corresponding dual SDP program, let us remind that the moment and localizing matrices $\mathbf{M}_{r-d_j}(g_j \, \mathbf{y})$ have entries which are linear in $\mathbf{y}$. Namely, one has $\mathbf{M}_{r-d_j}(g_j \, \mathbf{y}) = \sum_{\alpha \in \mathbb{N}_{2r}^n} \mathbf{C}_\alpha^j \, y_\alpha$; the matrix $\mathbf{C}_\alpha^j$ has rows and columns indexed by $\mathbb{N}_{r-d_j}^n$ with (β, γ)-entry equal to $\sum_{\beta+\gamma+\delta=\alpha} g_{j,\delta}$. In particular for $m = 0$, one has $g_0 = 1$ and the matrix $\mathbf{B}_\alpha := \mathbf{C}_\alpha^0$ has (β, γ)-entry equal to $1_{\beta+\gamma=\alpha}$, where $1_{\alpha=\beta}$ stands for the function which returns 1 if $\alpha = \beta$ and 0 otherwise. With $t_j := \binom{n+r-d_j}{r-d_j}$, the dual of SDP (2.7) is then the following SDP:

$$\begin{cases} \sup_{\mathbf{G}_j, b} & b \\ \text{s.t.} & f_\alpha - b1_{\alpha=0} = \sum_{j=0}^{m} \langle \mathbf{C}_\alpha^j, \mathbf{G}_j \rangle, \quad \alpha \in \mathbb{N}_{2r}^n \\ & \mathbf{G}_j \in \mathbb{S}_{t_j}^+, \quad j = 0, \ldots, m. \end{cases} \tag{2.8}$$

We can rewrite the equality constraints from SDP (2.8) in a more concise way, namely as $f - b \in \mathcal{M}(\mathfrak{g})_r$. To see this, let us first note that a sum of squares σ of degree $2r$ can be written as $\mathbf{v}^\top \mathbf{G} \mathbf{v}$, with

$$\mathbf{v} := (1, x_1, \ldots, x_n, x_1^2, x_1 x_2, \ldots, x_1^r, \ldots, x_n^r)$$

being the vectors of all monomials of degree at most r, and $\mathbf{G} \succeq 0$. The α-coefficient of $\sigma = \mathbf{v}^\top \mathbf{G} \mathbf{v}$ is equal to $\langle \mathbf{B}_\alpha, \mathbf{G} \rangle$. Similarly, for any $j \in [m]$ and SOS σ_j of degree at most $2(r - d_j)$, one can write $\sigma_j = \mathbf{v}_j^\top \mathbf{G}_j \mathbf{v}_j$, with $\mathbf{v}_j$ being the vector of all monomials of degree at most $r - d_j$, and $\mathbf{G}_j \succeq 0$. One can also check that the α-coefficient of $\sigma_j g_j$ is equal to $\langle \mathbf{C}_\alpha^j, \mathbf{G}_j \rangle$. Therefore, SDP (2.8) is equivalent to the following optimization problem over SOS polynomials:

$$\begin{cases} \sup\limits_{\sigma_j, b} & b \\ \text{s.t.} & f - b = \sum\limits_{j=0}^{m} \sigma_j g_j \\ & \sigma_j \in \Sigma[\mathbf{x}]_{r-d_j}, \quad j = 0, \ldots, m \end{cases} \tag{2.9}$$

or more concisely as

$$\sup\limits_{b} \{b : f - b \in \mathcal{M}(\mathfrak{g})_r\}. \tag{2.10}$$

The dual SDP (2.10) is obtained by replacing the nonnegativity condition $f - b \geq 0$ on $\mathbf{X}$ of the dual LP (2.6) by the more restrictive condition $f - b \in \mathcal{M}(\mathfrak{g})_r$. The sequences of SDP programs (2.7) and (2.10) are called the *moment* hierarchy and the SOS hierarchy, respectively. In the sequel, we refer to the sequence of primal-dual programs (2.7)–(2.10) as the moment-SOS hierarchy.

Theorem 2.5 *Under Assumption 2.2, the hierarchy of primal-dual moment-SOS relaxations (2.7)–(2.10) provides nondecreasing sequences of lower bounds converging to the global optimum $f_{\min}$ of* **P** *(2.4).*

The above theorem provides the theoretical convergence guarantee of the moment-SOS hierarchy.

Remark 2.6 *Even though we only included inequality constraints in the definition of* **X** *for the sake of simplicity, equality constraints can be treated in a dedicated way. For each equality constraint $h(\mathbf{x}) = 0$, with $h \in \mathbb{R}[\mathbf{x}]$, one adds the* *localizing constraint* $\mathbf{M}_{r-d_h}(h\,\mathbf{y}) = 0$, *with* $d_h := \lceil \deg(h)/2 \rceil$, *in the primal moment program* (2.7). *Similarly, in the dual SOS program* (2.9), *one adds a term* τh *to the sum* $\sum_{j=0}^{m} \sigma_j g_j$, *with* $\tau \in \mathbb{R}[\mathbf{x}]_{2r-2d_h}$.

In practice, it is possible to obtain finite convergence of the hierarchy, which is the topic of the next section.

2.4 Minimizer extraction

Here we describe sufficient conditions to obtain finite convergence of the moment-SOS hierarchy and extract the global minimizers of the polynomial f on $\mathbf{X}$.

Theorem 2.7 *Consider the sequence of primal moment relaxations defined in* (2.7). *If for some* $r \geq r_0 := \max\{d_1, \ldots, d_m\}$, *SDP* (2.7) *has an optimal solution* $\mathbf{y}$ *which satisfies*

$$\operatorname{rank} \mathbf{M}_{r'}(\mathbf{y}) = \operatorname{rank} \mathbf{M}_{r'-r_0}(\mathbf{y}) \text{ for some } r' \leq r, \qquad (2.11)$$

then $f^r = f_{\min}$ *and the infinite-dimensional LP* (2.5) *has an optimal solution* $\mu \in \mathcal{M}(\mathbf{X})_+$, *which is finitely supported on* $t = \operatorname{rank} \mathbf{M}_{r'}(\mathbf{y})$ *points of* $\mathbf{X}$, *or equivalently* t *global minimizers of* f *on* $\mathbf{X}$.

If the rank stabilization (also called *flatness*) condition (2.11) is satisfied, then finite convergence occurs, namely the SDP relaxation (2.7) is exact with optimal value $f^r = f_{\min}$. In addition, one can extract $\operatorname{rank} \mathbf{M}_{r'}(\mathbf{y})$ global minimizers of f on $\mathbf{X}$ with the following algorithm.

Algorithm 1 Extract

Require: The moment matrix $\mathbf{M}_{r'}(\mathbf{y})$ of rank t satisfying the flatness condition (2.11)
Ensure: The t points $\mathbf{x}(i) \in \mathbf{X}, i \in [t]$, global minimizers of Problem **P** (2.4)
1: Compute the Cholesky factorization $\mathbf{C}\mathbf{C}^{\mathsf{T}} = \mathbf{M}_{r'}(\mathbf{y})$
2: Reduce $\mathbf{C}$ to a column echelon form $\mathbf{U}$
3: Compute from $\mathbf{U}$ the multiplication matrices $\mathbf{N}_i, i \in [n]$
4: Compute $\mathbf{N} := \sum_{i=1}^{n} \lambda_i \mathbf{N}_i$ with randomly generated coefficients λ_i
5: Compute the Schur decomposition $\mathbf{N} = \mathbf{Q}\mathbf{T}\mathbf{Q}^{\mathsf{T}}$
6: Compute the column vectors $\{\mathbf{q}_j\}_{1 \leq j \leq t}$ of $\mathbf{Q}$
7: Return $x_i(j) := \mathbf{q}_j^{\mathsf{T}} \mathbf{N}_i \mathbf{q}_j, i \in [n], j \in [t]$

Proposition 2.8 *The procedure* Extract *described in Algorithm 1 is sound and returns t global optimizers of Problem* **P** *(2.4).*

PROOF Since the flatness condition (2.11) is satisfied, $\mathbf{y}$ is the moment sequence of a t-atomic Borel measure μ supported on $\mathbf{X}$. Namely, there are t points $\mathbf{x}(1), \dots, \mathbf{x}(t) \in \mathbf{X}$ such that

$$\mu = \sum_{j=1}^{t} \kappa_j \delta_{\mathbf{x}(j)}, \quad \kappa_j > 0, \quad \sum_{j=1}^{t} \kappa_j = 1.$$

By construction of the moment matrix $\mathbf{M}_{r'}(\mathbf{y})$, one has

$$\mathbf{M}_{r'}(\mathbf{y}) = \sum_{j=1}^{r} \kappa_j \mathbf{v}_{r'}(\mathbf{x}(j)) \mathbf{v}_{r'}^\mathsf{T}(\mathbf{x}(j)) = \mathbf{V} \mathbf{D} \mathbf{V}^\mathsf{T},$$

where the j-th column of $\mathbf{V}$ is $\mathbf{v}_{r'}(\mathbf{x}(j))$ and $\mathbf{D}$ is a $t \times t$ diagonal matrix with diagonal $(\kappa_j)_{1 \le j \le t}$. One can extract a Cholesky factor $\mathbf{C}$ as in Step 1, for instance via singular value decomposition. The following steps of the extraction algorithm consist of transforming $\mathbf{C}$ into $\mathbf{V}$ by suitable column operations. The reduction of $\mathbf{C}$ to a column echelon form in Step 2 is done by Gaussian elimination with column pivoting. By construction of the moment matrix, each row of $\mathbf{U}$ is indexed by a monomial $\mathbf{x}^\alpha$ involved in the vector $\mathbf{v}_{r'}$. Pivot elements in $\mathbf{U}$ correspond to monomials $\mathbf{x}^{\beta_j}$, $j \in [t]$ of the basis generating the t solutions. Namely, if $\mathbf{w} = (\mathbf{x}^{\beta_1}, \mathbf{x}^{\beta_2}, \dots, \mathbf{x}^{\beta_t})$ denotes this generating basis, then

$$\mathbf{v}_{r'}(\mathbf{x}(j)) = \mathbf{U} \mathbf{w}(\mathbf{x}(j)), \quad j \in [t].$$

Overall, extracting the global minimizers boils down to solving the above systems of equations. To solve this system, we compute at Step 3 each multiplication matrix $\mathbf{N}_i$, $i \in [n]$, which contains the coefficients of the monomials $x_i \mathbf{x}^{\beta_j}$, $j \in [t]$, namely which satisfy

$$\mathbf{N}_i \mathbf{w}(\mathbf{x}) = x_i \mathbf{w}(\mathbf{x}).$$

The entries of the global minimizers are all eigenvalues of the multiplication matrices. Since $\mathbf{w}(\mathbf{x}(j))$ is an eigenvector common to all multiplication matrices, one builds the random combination $\mathbf{N}$ of Step 4, which ensures with probability 1 that its eigenvalues are all distinct and have 1-dimensional eigenspaces. The Shur decomposition of Step 5 gives the decomposition $\mathbf{N} = \mathbf{Q} \mathbf{T} \mathbf{Q}^\mathsf{T}$ with an orthogonal matrix $\mathbf{Q}$ and an upper triangular matrix $\mathbf{T}$ with eigenvalues of $\mathbf{N}$ sorted in increasing order along the diagonal. □

Example 2.9 *Consider the polynomial optimization problem* **P** *(2.4) with* $f = -(x_1 - 1)^2 - (x_1 - x_2)^2 - (x_2 - 3)^2$ *and* $\mathbf{X} = \{\mathbf{x} \in \mathbb{R}^2 : 1 - (x_1 - 1)^2 \ge$

$0, 1 - (x_1 - x_2)^2 \geq 0, 1 - (x_1 - 3)^2 \geq 0$. *The first SDP relaxation outputs* $f^1 = -3$ *and* $\operatorname{rank} M_1(\mathbf{y}) = 3$, *while the second one outputs* $f^2 = -2$ *and the rank stabilizes with* $\operatorname{rank} M_1(\mathbf{y}) = \operatorname{rank} M_2(\mathbf{y}) = 3$. *Therefore the flatness condition holds, which implies that* $f_{\min} = f^2 = -2$. *The monomial basis is* $\mathbf{v}_2(\mathbf{x}) = (1, x_1, x_2, x_1^2, x_1 x_2, x_2^2)$. *The column echelon form* $\mathbf{U}$ *of the Cholesky factor of* $\mathbf{M}_2(\mathbf{y})$ *is given by*

$$\begin{bmatrix} 1 & & \\ 0 & 1 & \\ 0 & 0 & 1 \\ -2 & 3 & 0 \\ -4 & 2 & 2 \\ -6 & 0 & 5 \end{bmatrix}.$$

Pivot entries correspond to the generating basis $\mathbf{w}(\mathbf{x}) = (1, x_1, x_2)$. *Therefore the entries of the 3 global minimizers satisfy the following system of polynomial equations:*

$$x_1^2 = -2 + 3x_1$$
$$x_1 x_2 = -4 + 2x_1 + 2x_2$$
$$x_2^2 = -6 + 5x_2.$$

The multiplication matrices by x_1 *and* x_2 *can be extracted from rows in* $\mathbf{U}$ *as follows:*

$$\mathbf{N}_1 = \begin{bmatrix} 0 & 1 & 0 \\ -2 & 3 & 0 \\ -4 & 2 & 2 \end{bmatrix}, \quad \mathbf{N}_2 = \begin{bmatrix} 0 & 0 & 1 \\ -4 & 2 & 2 \\ -6 & 0 & 5 \end{bmatrix}.$$

After selecting a random convex combination of $\mathbf{N}_1$ *and* $\mathbf{N}_2$ *and computing the orthogonal matrix in the corresponding Schur decomposition, we obtain the 3 minimizers* $\mathbf{x}(1) = (1, 2)$, $\mathbf{x}(2) = (2, 2)$ *and* $\mathbf{x}(3) = (2, 3)$.

2.5 Notes and sources

The representation of positive polynomials stated in Theorem 2.3 is the well renowned Putinar's representation and is proved in [Put93]. Based on this representation, the convergence of the moment-SOS hierarchy, stated in Theorem 2.5, has been proved in [Las01].

The Riesz identification theorem can be found for instance in [LL01]. We refer the interested reader to [RF10, Section 21.7] and [Bar02, Chapter IV] or [Lue97, Section 5.10] for functional analysis, measure theory and applications in convex optimization.

The finite convergence of the moment-SOS hierarchy has been proved to hold generically under Assumption 2.2 in [Nie14]. The statement of Algorithm 1 extracting global minimizers and its correctness proof are available in [HL05] (combined with ideas from [LLR08]). The robustness of this algorithm has been studied in [KPV18]. It was proved that Algorithm 1 works under some generic conditions and Assumption 2.2 in [Nie13]. An interpretation of some wrong results, due to numerical inaccuracies, already observed when solving SDP relaxations for POP on a double precision floating point SDP solver is provided in [LM19].

Bibliography

[Bar02]   A. Barvinok. *A Course in Convexity*. Graduate studies in mathematics. American Mathematical Society, 2002.

[HL05]   D. Henrion and Jean-Bernard Lasserre. *Detecting Global Optimality and Extracting Solutions in GloptiPoly*, pages 293–310. Springer Berlin Heidelberg, Berlin, Heidelberg, 2005.

[KPV18]   Igor Klep, Janez Povh, and Jurij Volcic. Minimizer extraction in polynomial optimization is robust. *SIAM Journal on Optimization*, 28(4):3177–3207, 2018.

[Las01]   Jean-Bernard Lasserre. Global Optimization with Polynomials and the Problem of Moments. *SIAM Journal on Optimization*, 11(3):796–817, 2001.

[LL01]   Elliott H Lieb and Michael Loss. Analysis, graduate stud. math., vol. 14. In *Amer. Math. Soc*, 2001.

[LLR08]   Jean Bernard Lasserre, Monique Laurent, and Philipp Rostalski. Semidefinite characterization and computation of zero-dimensional real radical ideals. *Foundations of Computational Mathematics*, 8(5):607–647, 2008.

[LM19]   Jean-Bernard Lasserre and Victor Magron. In sdp relaxations, inaccurate solvers do robust optimization. *SIAM Journal on Optimization*, 29(3):2128–2145, 2019.

[Lue97]   D. G. Luenberger. *Optimization by Vector Space Methods*. John Wiley & Sons, Inc., New York, NY, USA, 1st edition, 1997.

[Nie13]   Jiawang Nie. Certifying convergence of Lasserre's hierarchy via flat truncation. *Mathematical Programming*, 142(1):485–510, 2013.

[Nie14]   Jiawang Nie. Optimality conditions and finite convergence of Lasserre's hierarchy. *Mathematical programming*, 146(1):97–121, 2014.

[Put93]   M. Putinar. Positive polynomials on compact semi-algebraic sets. *Indiana University Mathematics Journal*, 42(3):969–984, 1993.

[RF10]   H.L. Royden and P. Fitzpatrick. *Real Analysis*. Featured Titles for Real Analysis Series. Prentice Hall, 2010.

Part II

Correlative sparsity

Chapter 3

The moment-SOS hierarchy based on correlative sparsity

In this chapter, we describe how to exploit sparsity arising in the data of POPs from the perspective of variables, which leads to the notion of *correlative sparsity (CS)*. We start to explain how to build the correlative sparsity pattern (csp) graph in Section 3.1. Then, we provide in Section 3.2 an infinite-dimensional LP formulation over probability measures for POPs, based on CS. This LP program is then handled with a CS-adapted moment-SOS hierarchy of SDP relaxations, stated in Section 3.3. An alternative approach based on bounded degree SOS is given in Section 3.4. Section 3.5 focuses on minimizer extraction. Eventually, we explain in Section 3.6 how to extend this CS exploitation scheme to optimization over rational functions.

3.1 Correlative sparsity

Recall that a general POP is formulized as

$$\mathbf{P}: \quad f_{\min} = \inf_{\mathbf{x}} \{f(\mathbf{x}) : \mathbf{x} \in \mathbf{X}\}, \tag{3.1}$$

where $\mathbf{X} = \{\mathbf{x} \in \mathbb{R}^n : g_1(\mathbf{x}) \geq 0, \dots, g_m(\mathbf{x}) \geq 0\}$. Roughly speaking, the exploitation of CS in the moment-SOS hierarchy for $\mathbf{P}$ consists of two steps:

(1) decompose the variables $\mathbf{x}$ into a set of cliques according to the correlations between variables emerging in the input polynomial system;

(2) construct a sparse moment-SOS hierarchy with respect to the former decomposition of variables.

Let us proceed with more details. Recall $d_j := \lceil \deg(g_j)/2 \rceil$ for $j \in [m]$ and $r_{\min} := \max \{ \lceil \deg(f)/2 \rceil, d_1, \ldots, d_m \}$. Fix from now on a relaxation order $r \geq r_{\min}$. Let $J' := \{ j \in [m] \mid d_j = r \}$ which is possibly nonempty only when $r = r_{\min}$. We define the *csp* graph $G^{\mathrm{csp}}(V, E)$ associated to POP (3.1) whose node set is $V = \{1, 2, \ldots, n\}$ and whose edge set E satisfies $\{i, j\} \in E$ if one of following conditions holds:

(i) there exists $\alpha \in \mathrm{supp}(f) \cup \bigcup_{j \in J'} \mathrm{supp}(g_j)$ such that $\{i, j\} \subseteq \mathrm{supp}(\alpha)$;

(ii) there exists $k \in [m] \setminus J'$ such that $\{i, j\} \subseteq \bigcup_{\alpha \in \mathrm{supp}(g_k)} \mathrm{supp}(\alpha)$,

where $\mathrm{supp}(\alpha) := \{ k \in [n] \mid \alpha_k \neq 0 \}$ for $\alpha = (\alpha_1, \ldots, \alpha_n) \in \mathbb{N}^n$. Let $(G^{\mathrm{csp}})'$ be a chordal extension of G^{csp}[1] and $\{I_k\}_{k=1}^p$ be the list of maximal cliques of $(G^{\mathrm{csp}})'$ with $n_k := |I_k|$ so that the RIP (1.8) holds. Let $\mathbb{R}[\mathbf{x}, I_k]$ denote the ring of polynomials in the n_k variables $\{x_i\}_{i \in I_k}$ for $k \in [p]$. By construction, one can decompose the objective function f as $f = f_1 + \cdots + f_p$ with $f_k \in \mathbb{R}[\mathbf{x}, I_k]$ for all $k \in [p]$ (similarly for g_j with $j \in J'$). We then partition the constraint polynomials $g_j, j \in [m] \setminus J'$ into groups $\{ g_j \mid j \in J_k \}, k \in [p]$ which satisfy

(i) $J_1, \ldots, J_p \subseteq [m] \setminus J'$ are pairwise disjoint and $\bigcup_{k=1}^p J_k = [m] \setminus J'$;

(ii) for any $k \in [p]$ and any $j \in J_k$, $\bigcup_{\alpha \in \mathrm{supp}(g_j)} \mathrm{supp}(\alpha) \subseteq I_k$,

so that $g_j \in \mathbb{R}[\mathbf{x}, I_k]$ for all $k \in [p]$ and $j \in J_k$. In addition, suppose that Assumption 2.2 holds. Then all variables involved in POP (3.1) are bounded. To guarantee global convergence of the hierarchy that will be presented later, we need to add some redundant quadratic constraints to the description of the POP. We summarize all above in the following assumption.

Assumption 3.1 *Consider POP (3.1). The two index sets $[n]$ and $[m]$ are decomposed/partitioned into $\{I_1, \ldots, I_p\}$ and $\{J', J_1, \ldots, J_p\}$, respectively, such that*

(i) *The objective function f can be decomposed as $f = f_1 + \cdots + f_p$ with $f_k \in \mathbb{R}[\mathbf{x}, I_k]$ for $k \in [p]$ and the same goes for the constraint polynomial g_j with $j \in J'$;*

(ii) *For all $k \in [p]$ and $j \in J_k$, $g_j \in \mathbb{R}[\mathbf{x}, I_k]$;*

(iii) *The RIP (1.8) holds for $I_1, \ldots, I_p$ (possibly after some reordering);*

(iv) *For all $k \in [p]$, there exists $N_k > 0$ such that one of the constraint polynomials is $N_k - \sum_{i \in I_k} x_i^2$.*

---

[1]If G^{csp} is already a chordal graph, then we do not need the chordal extension.

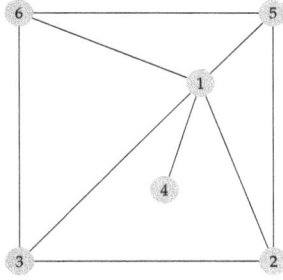

Figure 3.1: The csp graph for f over $\mathbf{X}$ from Example 3.9.

Example 3.2 *Consider an instance of POP (3.1) with* $f(\mathbf{x}) = x_2 x_5 + x_3 x_6 - x_2 x_3 - x_5 x_6 + x_1(-x_1 + x_2 + x_3 - x_4 + x_5 + x_6)$ *and*

$$\mathbf{X} = \{\mathbf{x} \in \mathbb{R}^n : g(\mathbf{x}) \geq 0, \text{ for all } g \in \mathfrak{g}\},$$

with $\mathfrak{g} = \{(6.36 - x_1)(x_1 - 4), \ldots, (6.36 - x_6)(x_6 - 4)\}$. *Here, there are* $n = 6$ *variables and the number of constraints is* $m = 6$. *The related csp graph* G^{csp} *is depicted in Figure 3.1. After adding an edge between nodes 3 and 5, the resulting graph* $(G^{\mathrm{csp}})'$ *is chordal with maximal cliques* $I_1 = \{1,4\}$, $I_2 = \{1,2,3,5\}$, $I_3 = \{1,3,5,6\}$. *Here* $p = 3$ *and one can write* $f = f_1 + f_2 + f_3$ *with*

$$f_1 = -x_1 x_4,$$
$$f_2 = -x_1^2 + x_1 x_2 + x_1 x_3 - x_2 x_3 + x_2 x_5,$$
$$f_3 = -x_5 x_6 + x_1 x_5 + x_1 x_6 + x_3 x_6.$$

For the relaxation order $r = r_{\min} = 1$, *let* $J' = [6]$ *and* $J_k = \varnothing$ *for* $k \in [3]$; *for the relaxation order* $r \geq 2$, *let* $J' = \varnothing$ *and* $J_1 = \{1,4\}$, $J_2 = \{2,3,5\}$, $J_3 = \{6\}$. *Then Assumption 3.1(i)–(ii) hold. In addition,* $I_1 \cap I_2 = \{1\} \subseteq I_3$, *and so RIP (1.8), or equivalently Assumption 3.1(iii), holds. For each* $i \in [n]$, *one has* $6.36^2 - x_i^2 \geq 0$ *for all* $\mathbf{x} \in \mathbf{X}$, *and so one can select* $N_1 = 2 \cdot 6.36^2$, $N_2 = N_3 = 4 \cdot 6.36^2$ *and add the redundant constraints* $N_k - \sum_{i \in I_k} x_i^2 \geq 0$, $k \in [p]$ *in the description of* $\mathbf{X}$, *so that Assumption 3.1(iv) holds as well.*

3.2 A sparse infinite-dimensional LP formulation

In this section, we assume $J' = \varnothing$. We now introduce a CS variant of the dense infinite-dimensional LP (2.5) formulation over probability measures stated in Section 2.3. The idea is to define a new measure for each subset I_k, $k \in [p]$, supported on a set $\mathbf{X}_k$ described by the constraints which only depend on the variables indexed by I_k, namely,

$$\mathbf{X}_k := \{\mathbf{x} \in \mathbb{R}^{n_k} : g_j(\mathbf{x}) \geq 0, j \in J_k\}, \text{ for } k \in [p].$$

So $\mathbf{X}$ can be equivalently described as

$$\mathbf{X} = \{\mathbf{x} \in \mathbb{R}^n : (x_i)_{i \in I_k} \in \mathbf{X}_k, k \in [p]\}. \tag{3.2}$$

Similarly, for all $j, k \in [p]$ such that $I_j \cap I_k \neq \varnothing$, define

$$\mathbf{X}_{jk} = \mathbf{X}_{kj} := \{(x_i)_{i \in I_j \cap I_k} : (x_i)_{i \in I_j} \in \mathbf{X}_j, (x_i)_{i \in I_k} \in \mathbf{X}_k\}.$$

Afterwards, for each $k \in [p]$ we define the projection $\pi_k : \mathcal{M}_+(\mathbf{X}) \rightarrow \mathcal{M}_+(\mathbf{X}_k)$ of the space of Borel measures supported on $\mathbf{X}$ on the space of Borel measures supported on $\mathbf{X}_k$, namely, for all $\mu \in \mathcal{M}_+(\mathbf{X})$,

$$\pi_k \mu(\mathbf{B}) := \mu(\{\mathbf{x} : \mathbf{x} \in \mathbf{X}, (x_i)_{i \in I_k} \in \mathbf{B}\}),$$

for each Borel set $\mathbf{B} \in \mathcal{B}(\mathbf{X}_k)$. We define similarly the projections π_{jk} for all $j, k \in [p]$ such that $I_j \cap I_k \neq \varnothing$. For each $k \in [p-1]$, we also rely on the set

$$U_k := \{j \in \{k+1, \ldots, p\} : I_j \cap I_k \neq \varnothing\}.$$

Then the CS variant of (2.5) reads as follows:

$$
\begin{aligned}
f_{\mathrm{cs}} := \quad &\inf_{\mu_k} \quad \sum_{k=1}^{p} \int_{\mathbf{X}_k} f_k(\mathbf{x}) \, \mathrm{d}\mu_k(\mathbf{x}) \\
&\text{s.t.} \quad \pi_{jk}\mu_j = \pi_{kj}\mu_k, \quad j \in U_k, k \in [p-1] \\
&\qquad \int_{\mathbf{X}_k} \mathrm{d}\mu_k(\mathbf{x}) = 1, \quad k \in [p] \\
&\qquad \mu_k \in \mathcal{M}_+(\mathbf{X}_k), \quad k \in [p].
\end{aligned}
\tag{3.3}
$$

To prove $f_{\mathrm{cs}} = f_{\min}$ under Assumption 3.1, we rely on the following auxiliary lemma, illustrated in Figure 3.2 in the case $p = 2$. This lemma uses the fact that one can disintegrate a probability measure on a product of Borel spaces into a marginal and a so-called *stochastic* kernel. Given two Borel spaces $\mathbf{X}, \mathbf{Z}$, a stochastic kernel $q(\mathrm{d}\mathbf{x}|\mathbf{z})$ on $\mathbf{X}$ given $\mathbf{Z}$ is defined by (1) $q(\mathrm{d}\mathbf{x}|\mathbf{z}) \in \mathcal{M}_+(\mathbf{X})$ for all $\mathbf{z} \in \mathbf{Z}$ and (2) the function $\mathbf{z} \mapsto q(\mathbf{B}|\mathbf{z})$ is $\mathcal{B}(\mathbf{Z})$-measurable for all $\mathbf{B} \in \mathcal{B}(\mathbf{Z})$.

Lemma 3.3 *Let* $[n] = \cup_{k=1}^{p} I_k$ *with* $n_k = |I_k|$, $\mathbf{X}_k \subseteq \mathbb{R}^{n_k}$ *be given compact sets, and let* $\mathbf{X} \subseteq \mathbb{R}^n$ *be defined as in* (3.2). *Let* $\mu_1 \in \mathcal{M}_+(\mathbf{X}_1), \ldots, \mu_p \in \mathcal{M}_+(\mathbf{X}_p)$ *be measures satisfying the equality constraints of LP* (3.3). *If RIP* (1.8) *holds for* $I_1, \ldots, I_p$, *then there exists a probability measure* $\mu \in \mathcal{M}_+(\mathbf{X})$ *such that*

$$\pi_k \mu = \mu_k, \tag{3.4}$$

for all $k \in [p]$, *that is,* μ_k *is the marginal of* μ *on* $\mathbb{R}^{n_k}$, *i.e., with respect to variables indexed by* I_k.

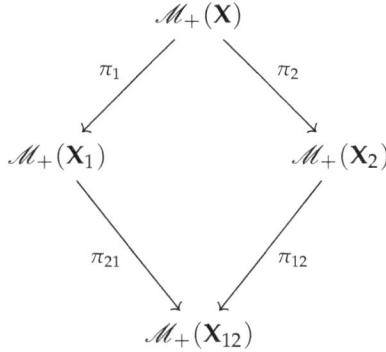

Figure 3.2: Illustration of Lemma 3.3 in the case $p = 2$.

PROOF The proof boils down to constructing μ by induction on p. If $p = 1$ and $I_1 = [n]$, the configuration corresponds exactly to the dense LP (2.5) formulation from Section 2.3, and one can simply take $\mu = \mu_1$. For the sake of conciseness, we only provide a proof for the case $p = 2$. Let $I_{12} := I_1 \cap I_2$ with cardinality n_{12}. If $I_{12} = \varnothing$, then one has $\mathbf{X} = \mathbf{X}_1 \times \mathbf{X}_2$ and we can simply define μ as the product measure of μ_1 and μ_2:

$$\mu(\mathbf{A} \times \mathbf{B}) := \mu_1(\mathbf{A}) \times \mu_2(\mathbf{B}),$$

for all $\mathbf{A} \in \mathcal{B}(\mathbb{R}^{n_1}), \mathbf{B} \in \mathcal{B}(\mathbb{R}^{n_2})$. This product measure μ satisfies (3.4).

Next, let us focus on the hardest case where $I_{12} \neq \varnothing$. Let $\overline{\pi}_k$ be the natural projection with respect to $I_k \backslash I_{12}$ and let us define the Borel set $\mathbf{Y}_k := \{\overline{\pi}_k(\mathbf{x}) : \mathbf{x} \in \mathbf{X}_k\} \in \mathcal{B}(\mathbb{R}^{n_k - n_{12}})$. It follows that μ_1, μ_2 can be seen as probability measures on the cartesian products $\mathbf{Y}_1 \times \mathbf{X}_{12} = \mathbf{X}_1$ and $\mathbf{X}_{12} \times \mathbf{Y}_2 = \mathbf{X}_2$, respectively. Let ν_1 and ν_2 be the stochastic kernels of μ_1 and μ_2, respectively. Since $\pi_{12}\mu_1 = \pi_{21}\mu_2 =: \nu$, one can disintegrate μ_1 and μ_2 as

$$\mu_1(\mathbf{A} \times \mathbf{B}) = \int_{\mathbf{B}} \nu_1(\mathbf{A}|\mathbf{x})\nu(\mathrm{d}\mathbf{x}), \quad \forall \mathbf{A} \in \mathcal{B}(\mathbf{Y}_1), \mathbf{B} \in \mathcal{B}(\mathbb{R}^{n_{12}}),$$

$$\mu_2(\mathbf{C} \times \mathbf{B}) = \int_{\mathbf{B}} \nu_2(\mathbf{C}|\mathbf{x})\nu(\mathrm{d}\mathbf{x}), \quad \forall \mathbf{A} \in \mathcal{B}(\mathbf{Y}_2), \mathbf{B} \in \mathcal{B}(\mathbb{R}^{n_{12}}).$$

Then, one can define the measure $\mu \in \mathcal{M}_+(\mathbf{Y}_1 \times \mathbb{R}^{n_{12}} \times \mathbf{Y}_2)$ as follows:

$$\mu(\mathbf{A} \times \mathbf{B} \times \mathbf{C}) = \int_{\mathbf{B}} \nu_1(\mathbf{A}|\mathbf{x})\nu_2(\mathbf{C}|\mathbf{x})\nu(\mathrm{d}\mathbf{x}),$$

for every Borel rectangle $\mathbf{A} \times \mathbf{B} \times \mathbf{C} \in \mathcal{B}(\mathbf{Y}_1) \times \mathcal{B}(\mathbf{X}_{12}) \times \mathcal{B}(\mathbf{Y}_2)$. In particular if $\mathbf{A} = \mathbf{Y}_1$, one has $\nu_1(\mathbf{A}|\mathbf{x}) = 1$ ν-a.e., and $\mu(\mathbf{Y}_1 \times \mathbf{B} \times \mathbf{C}) = \int_{\mathbf{B}} \nu_2(\mathbf{C}|\mathbf{x})\nu(\mathrm{d}\mathbf{x}) = \mu_2(\mathbf{B} \times \mathbf{C})$, implying that μ_2 is the marginal of μ on

$\mathbf{X}_{12} \times \mathbf{Y}_2 = \mathbf{X}_2$. Similarly, μ_1 is the marginal of μ on $\mathbf{Y}_1 \times \mathbf{X}_{12} = \mathbf{X}_1$, yielding the desired result. □

Now, we are ready to prove that LP (3.3) is not just a relaxation of the dense LP (2.5).

Theorem 3.4 *Consider POP* (3.1). *If Assumption 3.1 holds, then* $f_{cs} = f_{min}$.

PROOF The first inequality $f_{cs} \leq f_{min}$ is easy to show: let $\mathbf{a}$ be a global minimizer of f on $\mathbf{X}$, assumed to exist thanks to the compactness hypothesis. Let $\mu = \delta_{\mathbf{a}}$ be the Dirac measure concentrated on $\mathbf{a}$, and $\mu_k := \pi_k \mu$ be its projection on $\mathscr{M}_+(\mathbf{X}_k)$, for each $k \in [p]$. Namely, μ_k is the Dirac measure concentrated on $(a_i)_{i \in I_k} \in \mathbf{X}_k$, and is in particular a probability measure supported on $\mathbf{X}_k$. For each pair j, k such that $I_j \cap I_k \neq \emptyset$, the measure $\pi_{jk}\mu_j$ is the Dirac measure concentrated on $(a_i)_{i \in I_j \cap I_k} \in \mathbf{X}_{jk}$, and so is $\pi_{kj}\mu_k$. Therefore, each measure μ_k is a feasible solution of (3.3). In addition, the objective value of LP (3.3) is equal to $\sum_{k=1}^p f_k(\mathbf{a}) = f_{min}$, which proves the first inequality.

To prove the other inequality $f_{cs} \geq f_{min}$, let us fix a feasible solution (μ_k) of LP (3.3). By Lemma 3.3, there exists a probability measure $\mu \in \mathscr{M}_+(\mathbf{X})$ such that $\pi_k \mu = \mu_k$, for each $k \in [p]$. Then, one has

$$\sum_{k=1}^p \int_{\mathbf{X}_k} f_k \, d\mu_k = \sum_{k=1}^p \int_{\mathbf{X}_k} f_k \, d\mu = \int_{\mathbf{X}} \sum_{k=1}^p f_k \, d\mu = \int_{\mathbf{X}} f \, d\mu \geq f_{min}.$$

□

3.3 The CS-adpated moment-SOS hierarchy

In this section, we continue assuming $J' = \emptyset$. For $k \in [p]$, a moment sequence $\mathbf{y} \subseteq \mathbb{R}$ and $g \in \mathbb{R}[\mathbf{x}, I_k]$, let $\mathbf{M}_r(\mathbf{y}, I_k)$ (resp. $\mathbf{M}_r(g\,\mathbf{y}, I_k)$) be the moment (resp. localizing) submatrix obtained from $\mathbf{M}_r(\mathbf{y})$ (resp. $\mathbf{M}_r(g\,\mathbf{y})$) by retaining only those rows and columns indexed by $\beta \in \mathbb{N}_r^n$ of $\mathbf{M}_r(\mathbf{y})$ (resp. $\mathbf{M}_r(g\mathbf{y})$) with $\mathrm{supp}(\beta) \subseteq I_k$.

Example 3.5 *Consider again Example 3.2. The moment matrix* $\mathbf{M}_1(\mathbf{y}, I_1)$ *is indexed by the support vectors* $(0,0,0,0,0,0), (1,0,0,0,0,0), (0,0,0,1,0,0)$

(corresponding to the monomials 1, x_1 and x_4, respectively) and reads as follows:

$$\mathbf{M}_1(\mathbf{y}) = \left[\begin{array}{c|cc} 1 & y_{1,0,0,0,0,0} & y_{0,0,0,1,0,0} \\ \hline y_{1,0,0,0,0,0} & y_{2,0,0,0,0,0} & y_{1,0,0,1,0,0} \\ y_{0,0,0,1,0,0} & y_{1,0,0,1,0,0} & y_{0,0,0,2,0,0} \end{array} \right].$$

With $r \geq r_{\min}$, the moment hierarchy based on CS for POP (3.3) is defined as

$$\begin{cases} \inf\limits_{\mathbf{y}_k} & \sum_{k=1}^p L_{\mathbf{y}_k}(f_k) \\ \text{s.t.} & \mathbf{M}_r(\mathbf{y}_k, I_k) \succeq 0, \quad k \in [p] \\ & \mathbf{M}_{r-d_j}(g_j\mathbf{y}_k, I_k) \succeq 0, \quad j \in J_k, k \in [p] \\ & L_{\mathbf{y}_k}(\mathbf{x}^\alpha) = L_{\mathbf{y}_j}(\mathbf{x}^\alpha), \alpha \in \mathbb{N}_{2r}^n, \operatorname{supp}(\alpha) \subseteq I_k \cap I_j, j \in U_k, k \in [p] \\ & L_{\mathbf{y}_k}(1) = 1, \quad k \in [p]. \end{cases}$$

(3.5)

Note that SDP (3.5) is equivalent to the following program:

$$\mathbf{P}_{cs}^r: \begin{cases} \inf\limits_{\mathbf{y}} & L_{\mathbf{y}}(f) \\ \text{s.t.} & \mathbf{M}_r(\mathbf{y}, I_k) \succeq 0, \quad k \in [p] \\ & \mathbf{M}_{r-d_j}(g_j\mathbf{y}, I_k) \succeq 0, \quad j \in J_k, k \in [p] \\ & y_0 = 1 \end{cases}$$

(3.6)

with optimal value denoted by f_{cs}^r. Indeed, for a sequence $\mathbf{y} = (y_\alpha)_{\alpha \in \mathbb{N}_{2r}^n}$, one can define $\mathbf{y}_k := \{y_\alpha : \alpha \in \mathbb{N}_{2r}^n, \operatorname{supp}(\alpha) \subseteq I_k\}$, for all $k \in [p]$. Cne obviously has $L_{\mathbf{y}_k}(1) = 1$, and each moment matrix $\mathbf{M}_r(\mathbf{y}_k, I_k)$ is equal to $\mathbf{M}_r(\mathbf{y}, I_k)$ (and similarly for the localizing matrices). In addition. if $I_k \cap I_j \neq \varnothing$ and $\operatorname{supp}(\alpha) \subseteq I_k \cap I_j$, then

$$L_{\mathbf{y}_k}(\mathbf{x}^\alpha) = \{y_\alpha : \operatorname{supp}(\alpha) \subseteq I_k \cap I_j\} = L_{\mathbf{y}_j}(\mathbf{x}^\alpha).$$

Let $\Sigma[\mathbf{x}, I_k] \subseteq \mathbb{R}[\mathbf{x}, I_k]$ be the corresponding cone of SOS polynomials. Then the dual of (3.6) is

$$\begin{cases} \sup\limits_{b, \sigma_{k,j}} & b \\ \text{s.t.} & f - b = \sum_{k=1}^p \left(\sigma_{k,0} + \sum_{j \in J_k} \sigma_{k,j} g_j \right) \\ & \sigma_{k,0}, \sigma_{k,j} \in \Sigma[\mathbf{x}, I_k], \quad j \in J_k, k \in [p] \\ & \deg(\sigma_{k,0}), \deg(\sigma_{k,j} g_j) \leq 2r, \quad j \in J_k, k \in [p]. \end{cases}$$

(3.7)

In the following, we refer to (3.6)–(3.7) as the CS-adpated moment-SOS (CSSOS) hierarchy. To prove that the sequence $(f_{cs}^r)_{r \geq r_{\min}}$ converges to the global optimum $f_{\min}$ of the original POP (3.1), we rely on Lemma 3.3.

> **Theorem 3.6** *Consider POP* (3.1). *If Assumption 3.1 holds, then the CSSOS hierarchy* (3.6)–(3.7) *provides a nondecreasing sequence of lower bounds* $(f^r_{cs})_{r \geq r_{min}}$ *converging to* f_{min}.

Remark 3.7 *Despite the convergence guarantee stated in Theorem 3.6, note that SDP* (3.6) *is a relaxation of the dense SDP* (2.7) *in general, and one can have* $f^r_{cs} < f^r$ *for some relaxation order r. The underlying reason is that the situation here is different from the case of PSD matrix completion (Theorem 1.5). Namely, there is no guarantee that one can obtain a PSD matrix completion* $\mathbf{M}_r(\mathbf{y})$ *from the submatrices* $\mathbf{M}_r(\mathbf{y}, I_k)$, $k \in [p]$ *because of the specific Hankel structure of* $\mathbf{M}_r(\mathbf{y})$. *At the end of Appendix B.1, we provide a Julia script showing such conservatism behavior.*

As a corollary of Theorem 3.6, we obtain the following representation result, which is a CS version of Putinar's Positivstellensatz.

> **Theorem 3.8** *Let* $f \in \mathbb{R}[\mathbf{x}]$ *be positive on a basic compact semialgebraic set* **X** *as in* (2.1). *Let Assumption 3.1 hold. Then,*
>
> $$f = \sum_{k=1}^{p} \left(\sigma_{k,0} + \sum_{j \in J_k} \sigma_{k,j} g_j \right),\tag{3.8}$$
>
> *for some polynomials* $\sigma_{k,0}, \sigma_{k,j} \in \Sigma[\mathbf{x}, I_k]$, $j \in J_k$, $k \in [p]$.

Let us compare the computational cost of the CSSOS hierarchy (3.7) with the dense hierarchy (2.9). For this, we define $\tau := \max_{k \in [p]} |I_k| = \max_{k \in [p]} n_k$, that is, τ is the maximal size of the subsets $I_1, \ldots, I_p$.

(1) The dense SOS formulation (2.9) involves $m + 1$ SOS polynomials in n variables of degree at most $2r$, yielding $m + 1$ SDP matrices of size at most $\binom{n+r}{r}$ and $\binom{n+2r}{2r}$ equality constraints.

(2) The CSSOS formulation (3.7) involves $p + m$ SOS polynomials in at most τ variables and of degree at most $2r$, yielding $p + m$ SDP matrices of size at most $\binom{\tau+r}{r}$ and at most $p\binom{\tau+2r}{2r}$ equality constraints.

Overall, when n is fixed and r varies, the r-th step of the hierarchy involves $\mathcal{O}(r^{2n})$ equality constraints in the dense setting against $\mathcal{O}(pr^{2\tau})$ in the sparse setting. This allows one to handle POPs involving several hundred variables if the maximal subset size τ is small (say, $\tau \leq 10$). Furthermore, as shown in the following example, one can also benefit from the computational cost saving when r increases for POPs involving a small number of variables (say, $n \leq 10$).

Example 3.9 *Coming back to Example 3.2, let us compare the hierarchy of dense relaxations given in Section 2.3 with the CS variant. For $r = 1$, the dense SDP relaxation (2.9) involves $\binom{n+2r}{2r} = \binom{6+2}{2} = 28$ equality constraints and provides a lower bound of $f^1 = 20.755$ for $f_{\min}$. The dense SDP relaxation (2.9) with $r = 2$ involves $\binom{6+4}{4} = 210$ equality constraints and provides a tighter lower bound of $f^2 = 20.8608$. For $r = 2$, the sparse SDP relaxation (3.7) involves $\binom{2+4}{4} + 2\binom{4+4}{4} = 155$ equality constraints and provides the same bound $f_{cs}^2 = f^2 = 20.8608$. In Appendix A, we provide the MATLAB script allowing one to obtain these results. The dense SDP relaxation with $r = 3$ involves 924 equality constraints against 448 for the sparse variant. This difference becomes significant while considering the polynomial time complexity of solving SDP, already mentioned in Section 1.3.*

3.4 A variant with SOS of bounded degrees

In certain situations, alternative representations for positive polynomials can be more interesting as one can bound in advance the degree of the SOS polynomials involved.

Theorem 3.10 *Let us fix an $r \in \mathbb{N}$ and let Assumption 3.1 hold with $J' = \emptyset$. Suppose that, possibly after scaling, $0 \leq g_j \leq 1$ on $\mathbf{X}$ for each $j \in [m]$. If f is positive on $\mathbf{X}$, then*

$$f = \sum_{k=1}^{p} \left(\sigma_k + \sum_{\alpha, \beta \in \mathbb{N}^{|J_k|}} c_{k,\alpha\beta} \prod_{j \in J_k} g_j^{\alpha_j} (1 - g_j)^{\beta_j} \right), \qquad (3.9)$$

for some finitely many real scalars $c_{k,\alpha\beta} \geq 0$ and SOS polynomials $\sigma_k \in \Sigma[\mathbf{x}, I_k]$ with $\deg(\sigma_k) \leq 2r$.

The representation from Theorem 3.10 is called the *sparse bounded SOS (SBSOS) representation*. After replacing the right-hand side in the equality constraint of (3.7) by an SBSOS representation, we obtain the following

SBSOS hierarchy, indexed by an integer $s \in \mathbb{N}^*$:

$$\begin{cases} \sup_{b,\sigma_k,c_{k,\alpha\beta}} & b \\ \text{s.t.} & f - b = \sum_{k=1}^p \left(\sigma_k + \sum_{|\alpha+\beta| \leq s} c_{k,\alpha\beta} \prod_{j \in J_k} g_j^{\alpha_j} (1 - g_j)^{\beta_j} \right) \quad (3.10) \\ & c_{k,\alpha\beta} \geq 0, |\alpha + \beta| \leq s, \quad k \in [p], \alpha, \beta \in \mathbb{N}^{|J_k|} \\ & \sigma_k \in \Sigma[\mathbf{x}, I_k], \deg(\sigma_k) \leq 2r, \quad j \in J_k, k \in [p]. \end{cases}$$

While the degree of each σ_k is a priori fixed and at most $2r$, one can increase the degree s of each term $\prod_{j \in J_k} g_j^{\alpha_j} (1 - g_j)^{\beta_j}$, which boils down to multiplying the polynomials describing the constraint set $\mathbf{X}$.

Regarding computational costs, SDP (3.10) involves p LMI constraints (associated to the σ_k) of maximal size $\binom{\tau+r}{r}$ (recall $\tau = \max_{k \in [p]} |I_k|$), which does not depend on s. In addition, the number of coefficients $c_{k,\alpha\beta}$ is equal to $\binom{|J_k|+s}{s}$. Therefore, the SBSOS hierarchy offers a computational benefit when each $|J_k|$ is relatively small.

3.5 Minimizer extraction

As for the standard dense moment-SOS hierarchy stated in Section 2.3, one can also detect finite convergence of the CSSOS hierarchy and extract global minimizers with a dedicated extraction algorithm — the CS variant of Algorithm 1.

Theorem 3.11 *Consider POP (3.1). Let Assumption 3.1(i)–(ii) hold and let us consider the hierarchy of moment relaxations $(\mathbf{P}_{cs}^r)_{r \geq r_{\min}}$ defined in (3.6). Let $a_k := \max_{j \in J_k}\{d_j\}$ for all $k \in [p]$. If for some $r \geq r_{\min}$, $\mathbf{P}_{cs}^r$ has an optimal solution $\mathbf{y}$ which satisfies*

$$\text{rank}\, \mathbf{M}_r(\mathbf{y}, I_k) = \text{rank}\, \mathbf{M}_{r-a_k}(\mathbf{y}, I_k) \text{ for all } k \in [p], \quad (3.11)$$

and $\text{rank}\, \mathbf{M}_r(\mathbf{y}, I_j \cap I_k) = 1$ *for all pairs* (j,k) *with* $I_j \cap I_k \neq \emptyset$*, then the SDP relaxation (3.6) is exact, i.e., $f_{cs}^r = f_{\min}$. In addition, for each $k \in [p]$, let $\Delta_k := \{\mathbf{x}(k)\} \subseteq \mathbb{R}^{n_k}$ be a set of solutions obtained from the extraction procedure* Extract, *stated in Algorithm 1 and applied to the moment matrix $\mathbf{M}_r(\mathbf{y}, I_k)$. Then every $\mathbf{x} \in \mathbb{R}^n$ obtained by $(x_i)_{i \in I_k} = \mathbf{x}(k)$ for some $\mathbf{x}(k) \in \Delta_k$ is a global minimizer of POP (3.1).*

Note that Assumption 3.1(iii)–(iv) are not required in Theorem 3.11, as the rank conditions are strong enough to ensure finite convergence and extraction of a subset of global minimizers.

3.6 From polynomial to rational functions

Here, we consider the following optimization problem:

$$f_{\min} := \inf_{x \in X} \sum_{i=1}^{t} \frac{p_i(x)}{q_i(x)}, \tag{3.12}$$

where $X = \{x \in \mathbb{R}^n : g_1(x) \geq 0, \ldots, g_m(x) \geq 0\}$ is a compact set, all numerators and denominators are polynomials, and all denominators are positive on X.

Problem (3.12) is a *fractional programming* problem of a rather general form. Here, we assume that the degree of each numerator/denominator is rather small (≤ 10), but that the number of terms t can be much larger (10 to 100). For dense data, the number of variables should also be small (≤ 10). However, this number can be also relatively large (10 to 100) provided that the problem data exhibits CS, as in Section 3.1. One naive strategy is to reduce all fractions to the same denominator and obtain a single rational fraction to minimize. However, this approach is not adequate since the degree of the common denominator will be potentially large and even if n is small, one might not be able to solve the first-order relaxation of the related moment-SOS hierarchy. A more suitable strategy for solving (3.12) is to cast it as a particular instance of the GMP, namely the following infinite-dimensional LP over measures:

$$f_{\text{meas}} := \inf_{\mu_i} \sum_{i=1}^{t} \int_X p_i(x)\, d\mu_i(x)$$

$$\text{s.t.} \int_X x^\alpha q_i(x) d\mu_i(x) = \int_X x^\alpha q_1(x) d\mu_1(x), \alpha \in \mathbb{N}^n, i = 2, \ldots, t$$

$$\int_X q_1(x)\, d\mu_1(x) = 1$$

$$\mu_1, \ldots, \mu_t \in \mathcal{M}_+(X). \tag{3.13}$$

Theorem 3.12 *Consider* (3.12). *Let* X *be a compact set and assume that each q_i is positive on X for all $i \in [t]$. Then $f_{\text{meas}} = f_{\min}$.*

PROOF First, we prove that $f_{\min} \geq f_{\text{meas}}$. Let a be a global minimizer of $f = \sum_{i=1}^{t} \frac{p_i}{q_i}$ on X, assumed to exist thanks to the compactness hypothesis. For each $i \in [t]$, let $\mu_i = \frac{1}{q_i(a)} \delta_a$ be the Dirac measure centered at

a weighted by $\frac{1}{q_i(\mathbf{a})}$. We then have

$$\int_{\mathbf{X}} q_1(\mathbf{x}) \, d\mu_1(\mathbf{x}) = \frac{1}{q_1(\mathbf{a})} \int_{\mathbf{X}} q_1(\mathbf{x}) \delta_{\mathbf{a}} \, d\mathbf{x} = 1,$$

and for each $i < t$ and all $\alpha \in \mathbb{N}^n$,

$$\int_{\mathbf{X}} q_i(\mathbf{x}) \mathbf{x}^\alpha \, d\mu_i(\mathbf{x}) = \mathbf{a}^\alpha,$$

ensuring that the measures $(\mu_i)_{i \in [t]}$ are feasible for (3.13). The associated optimal value is

$$\sum_{i=1}^{t} \int_{\mathbf{X}} p_i(\mathbf{x}) \, d\mu_i(\mathbf{x}) = \sum_{i=1}^{t} \frac{p_i(\mathbf{a})}{q_i(\mathbf{a})} = f(\mathbf{a}) = f_{\min}.$$

To prove the other direction, let $(\mu_i)_{i \in [t]}$ be a feasible solution of (3.13). For each $i \in [t]$, let us define the measure ν_i as follows:

$$\nu_i(\mathbf{B}) := \int_{\mathbf{X} \cap \mathbf{B}} q_i(\mathbf{x}) \, d\mu_i(\mathbf{x}),$$

for each Borel set $\mathbf{B}$ in the Borel σ-algebra of $\mathbb{R}^n$. The support of ν_i is $\mathbf{X}$. Since measures on compact sets are moment determinate (see Section 2.2), the moment equality constraints of (3.13) imply that $\nu_i = \nu$, for each $i \in \{2, \ldots, t\}$. The constraint $\int_{\mathbf{X}} q_1 \, d\mu_1 = 1$ implies that ν_1 is a probability measure on $\mathbf{X}$ (since its mass is 1). Then one can rewrite the objective value as

$$\sum_{i=1}^{t} \int_{\mathbf{X}} p_i \, d\mu_i = \sum_{i=1}^{t} \int_{\mathbf{X}} \frac{p_i}{q_i} q_i \, d\mu_i = \sum_{i=1}^{t} \int_{\mathbf{X}} \frac{p_i}{q_i} \, d\nu_1$$

$$= \int_{\mathbf{X}} f \, d\nu_1 \geq \int_{\mathbf{X}} f_{\min} \, d\nu_1 = f_{\min},$$

where the inequality follows from the fact that $f \geq f_{\min}$ on $\mathbf{X}$. $\square$

This first GMP formulation (3.13) allows one to handle general (possibly dense) rational programs. When the numerators and denominators satisfy a csp similar to the one stated in Section 3.1, one can rely on a dedicated CS formulation.

Assumption 3.13 *There is a decomposition of $[n] = \cup_{i=1}^{t} I_i$ and a partition of $[m] = \cup_{i=1}^{t} J_i$ such that for each $i \in [t]$, $p_i, q_i \in \mathbb{R}[\mathbf{x}, I_i]$ (as in Assumption 3.1(i) in the case of polynomials), and Assumption 3.1(ii)–(iv) hold.*

Then, as in Section 3.2 for each $i \in [t]$, with $n_i = |I_i|$, let us define

$$X_i := \{x \in \mathbb{R}^{n_i} : g_j(x) \geq 0, j \in J_i\},$$

so that X can be equivalently described as

$$X = \{x \in \mathbb{R}^n : (x_k)_{k \in I_i} \in X_i, i \in [t]\}.$$

Similarly, for all pairs $i, j \in [t]$ such that $I_i \cap I_j \neq \varnothing$, we define X_{ij}, as well as the projection $\pi_i : \mathcal{M}_+(X) \to \mathcal{M}_+(X_i)$, for each $i \in [t]$, and π_{ij} for all pairs $i, j \in [t]$ such that $I_i \cap I_j \neq \varnothing$. Then the CS variant of the infinite-dimensional LP (3.13) is given by

$$
\begin{aligned}
f_{cs} := \quad &\inf_{\mu_i} \quad \sum_{i=1}^{t} \int_{X_i} p_i(x)\, d\mu_i(x) \\
&\text{s.t.} \quad \pi_{ij}(q_i\, d\mu_i) = \pi_{ji}(q_j\, d\mu_j), \quad j \in U_i, i \in [t-1] \qquad (3.14) \\
&\qquad \int_{X_i} q_i(x)\, d\mu_i(x) = 1, \mu_i \in \mathcal{M}_+(X_i), \quad i \in [t].
\end{aligned}
$$

Theorem 3.14 *Consider (3.12). Let X be a compact set, and assume that q_i is positive on X for all $i \in [t]$. If Assumption 3.13 holds, then $f_{cs} = f_{\min}$.*

One can then derive a CS-adpated hierarchy of SDP relaxations for LP (3.14):

$$
\mathbf{P}_{cs}^r :
\begin{cases}
\inf_{y_i} \quad \sum_{k=1}^{t} L_{y_i}(p_i) \\
\text{s.t.} \quad M_r(y_i, I_i) \succeq 0, \quad i \in [t] \\
\qquad M_{r-d_j}(g_j y_i, I_i) \succeq 0, \quad j \in J_i, i \in [t] \\
\qquad L_{y_i}(x^\alpha q_i) = L_{y_j}(x^\alpha q_j), \quad |\alpha| + \max\{\deg(q_i), \deg(q_j)\} \leq 2r \\
\qquad\qquad\qquad\qquad\qquad\qquad\quad \text{and } \operatorname{supp}(\alpha) \subseteq I_i \cap I_j, j \in U_i, i \in [t] \\
\qquad L_{y_i}(q_i) = 1, \quad i \in [t]
\end{cases}
$$
$$(3.15)$$

where $|\alpha| := \sum_{i=1}^{n} \alpha_i$ for $\alpha \in \mathbb{N}^n$.

Example 3.15 *From the Rosenbrock problem*

$$\inf_{x \in \mathbb{R}^n} \sum_{i=1}^{n-1} \left(100(x_{i+1} - x_i^2)^2 + (x_i - 1)^2\right),$$

we define the following rational optimization problem

$$f_{\max} := \sup_{x \in \mathbb{R}^n} \sum_{i=1}^{n-1} \frac{1}{100(x_{i+1} - x_i^2)^2 + (x_i - 1)^2}. \tag{3.16}$$

Note that Problem (3.16) has the same critical points as the Rosenbrock problem, which yields numerical issues for local optimization solvers embedded in optimization software such as BARON or NEOS. The global optimum $f_{\max} = n - 1$ of Problem (3.16) is attained at $x_i = 1$, $i \in [n]$. Each summand depends on 2 variables, so that we can define $I_i = \{i, i+1\}$ for $i \in [n-1]$. To bound the problem, we let $g_i = 16 - x_i^2$ for $i \in [n]$. When calling multiple times local optimization solvers with random initial guesses, we obtain local optima far away from the global optimum. This is in deep contrast with our CS-adapted hierarchy of SDP (3.15). After selecting the minimal relaxation order $r = 1$, we obtain the global optimum for Problem (3.16) with up to 1000 variables. The typical CPU time ranges from a few seconds to a few minutes on a modern standard laptop. The BARON software (on the NEOS server) can find the global optimum in most cases when $n < 640$. For $n \geq 640$, BARON returns wrong values (the first corresponding coordinate is equal to -0.995 instead of 1). Overall this rational problem yields numerical instabilities for such local solvers, which can be handled with the CS-adapted SDP relaxations.

3.7   Notes and sources

The CSSOS hierarchy for POPs was first studied in [WKKM06]. Its global convergence was proved in [Las06] soon after by introducing p additional quadratic constraints. Lemma 3.3 is proved in [Las06, §6]. Theorem 3.8 is stated and proved in [Las06, Corollary 3.9]. An alternative proof is provided in [GNS07].

The alternative SBSOS representation stated in Theorem 3.10 is provided in [WLT18]. We refer to this former paper for more details on properties of the related SOS hierarchy.

Theorem 3.11 is stated in [Las06, Theorem 3.7] and proved in [Las06, §4.2].

The results from Section 3.6 are mostly stated in [BHL16]. The proof of Theorem 3.14, very similar to the one of Theorem 3.4, can be found in [BHL16, §3.1]. Example 3.15 is provided in [BHL16, §4.4.2]. The interested reader can find detailed illustrations of the CSSOS hierarchy for rational programming in [BHL16, §4], together with comparison with local optimization solvers such as the BARON software [TS05], publicly available on the NEOS server [CMM98].

For the more advanced problem of set approximation, in particular in the context of dynamical systems, exploiting CS is more delicate as the set of trajectories of a system with sparse dynamics is not necessarily sparse. Recent research efforts have been pursued for volume approximation [TWLH21] and region of attraction [TCHL20, SK20].

Bibliography

[BHL16]   Florian Bugarin, Didier Henrion, and Jean Bernard Lasserre.
            Minimizing the sum of many rational functions. *Mathematical
            Programming Computation*, 8(1):83–111, 2016.

[CMM98]   Joseph Czyzyk, Michael P Mesnier, and Jorge J Moré. The
            neos server. *IEEE Computational Science and Engineering*,
            5(3):68–75, 1998.

[GNS07]   David Grimm, Tim Netzer, and Markus Schweighofer. A note
            on the representation of positive polynomials with structured
            sparsity. *Archiv der Mathematik*, 89(5):399–403, 2007.

[Las06]   Jean B Lasserre. Convergent sdp-relaxations in polynomial
            optimization with sparsity. *SIAM Journal on Optimization*,
            17(3):822–843, 2006.

[SK20]   Corbinian Schlosser and Milan Korda. Sparse moment-sum-
            of-squares relaxations for nonlinear dynamical systems with
            guaranteed convergence. *arXiv preprint arXiv:2012.05572*,
            2020.

[TCHL20]   Matteo Tacchi, Carmen Cardozo, Didier Henrion, and
            Jean Bernard Lasserre. Approximating regions of attraction of
            a sparse polynomial differential system. *IFAC-PapersOnLine*,
            53(2):3266–3271, 2020.

[TS05]   Mohit Tawarmalani and Nikolaos V Sahinidis. A polyhedral
            branch-and-cut approach to global optimization. *Mathemati-
            cal programming*, 103(2):225–249, 2005.

[TWLH21]   Matteo Tacchi, Tillmann Weisser, Jean Bernard Lasserre, and
            Didier Henrion. Exploiting sparsity for semi-algebraic set vol-
            ume computation. *Foundations of Computational Mathematics*,
            pages 1–49, 2021.

[WKKM06] Hayato Waki, Sunyoung Kim, Masakazu Kojima, and Masakazu Muramatsu. Sums of squares and semidefinite program relaxations for polynomial optimization problems with structured sparsity. *SIAM Journal on Optimization*, 17(1):218–242, 2006.

[WLT18]   Tillmann Weisser, Jean B Lasserre, and Kim-Chuan Toh. Sparse-bsos: a bounded degree sos hierarchy for large scale polynomial optimization with sparsity. *Mathematical Programming Computation*, 10(1):1–32, 2018.

Chapter 4

Application in computer arithmetic

In this chapter, we describe an optimization framework to provide upper bounds on absolute roundoff errors of floating-point nonlinear programs, involving polynomials. The efficiency of this framework is based on the CSSOS hierarchy which exploits CS of the input polynomials, as described in Chapter 3.

Constructing numerical programs which perform accurate computation turns out to be difficult, due to finite numerical precision of implementations such as floating-point or fixed-point representations. Finite-precision numbers induce roundoff errors, and knowledge of the range of these roundoff errors is required to fulfill safety criteria of critical programs, as typically arising in modern embedded systems such as aircraft controllers. Such knowledge can be used in general for developing accurate numerical software, but is also particularly relevant when considering migration of algorithms onto hardware (e.g., FPGAs). The advantage of architectures based on FPGAs is that they allow more flexible choices in number representations, rather than limiting the choice between IEEE standard single or double precision. Indeed, in this case, we can benefit from a more flexible number representation while still ensuring guaranteed bounds on the program output.

Our method to bound the error is a decision procedure based on a specialized variant of the Lasserre hierarchy [Las06], outlined in Chapter 3. The procedure relies on SDP to provide sparse SOS decompositions of nonnegative polynomials. Our framework handles polynomial program analysis (involving the operations $+, \times, -$) as well as extensions to the more general class of semialgebraic and transcendental programs (involving $\sqrt{}, /, \min, \max, \arctan, \exp$), following the approximation scheme described in [MAGW15]. For the sake of conciseness, we focus in this book

47

on polynomial programs only. The interested reader can find more details in the related publication [MCD17].

4.1 Polynomial programs

Here we consider a given program that implements a polynomial expression f with input variables x taking values in a region X. We assume that X is included in a box (i.e., a product of closed intervals) and that X is encoded as in (2.1):

$$X := \{\, x \in \mathbb{R}^n \; : \; g_1(x) \geq 0, \ldots, g_l(x) \geq 0 \,\},$$

for polynomial functions $g_1, \ldots, g_l$.

The type of numerical constants is denoted by C. In our current implementation, the user can choose either 64 bit floating-point or arbitrary-size rational numbers. The inductive type of polynomial expressions $f, g_1, \ldots, g_l$ with coefficients in C is pExprC defined as follows:

```
type pexprC =
  Pc of C
| Px of positive
| Psub of pexprC * pexprC | Pneg of pexprC
| Padd of pexprC * pexprC
| Pmul of pexprC * pexprC
```

The constructor Px takes a positive integer as argument to represent either an input or local variable. One obtains rounded expressions using a recursive procedure round. We adopt the standard practice [Hig02] to approximate a real number x with its closest floating-point representation $\hat{x} = x(1+e)$, with $|e|$ is less than the machine precision ε. In the sequel, we neglect both overflow and denormal range values. The operator $\hat{\cdot}$ is called the rounding operator and can be selected among rounding to nearest, rounding toward zero (resp. $\pm\infty$). In the sequel, we assume rounding to nearest. The scientific notation of a binary (resp. decimal) floating-point number $\hat{x}$ is a triple (s, sig, exp) consisting of a sign bit s, a *significand* $sig \in [1, 2)$ (resp. $[1, 10)$) and an *exponent* exp, yielding numerical evaluation $(-1)^s \, sig \, 2^{exp}$ (resp. $(-1)^s \, sig \, 10^{exp}$).

The upper bound on the relative floating-point error is given by $\varepsilon = 2^{-prec}$, where prec is called the *precision*, referring to the number of significand bits used. For single precision floating-point, one has prec $= 24$. For double (resp. quadruple) precision, one has prec $= 53$ (resp. prec $= 113$). Let F be the set of binary floating-point numbers.

For each real-valued operation $bop_{\mathbb{R}} \in \{+, -, \times\}$, the result of the corresponding floating-point operation $bop_F \in \{\oplus, \ominus, \otimes\}$ satisfies the following when complying with IEEE 754 standard arithmetic [IEE08] (without overflow, underflow and denormal occurrences):

$$\mathrm{bop_F}(\hat{x}, \hat{y}) = \mathrm{bop_R}(\hat{x}, \hat{y})(1 + e), \quad |e| \leq \varepsilon = 2^{-\mathrm{prec}}. \tag{4.1}$$

Then, we denote by $\hat{f}(\mathbf{x}, \mathbf{e})$ the rounded expression of f after applying the round procedure, introducing additional error variables $\mathbf{e}$.

4.2 Upper bounds on roundoff errors

The algorithm `roundoff_bound`, depicted in Algorithm 2, takes as input $\mathbf{x}$, $\mathbf{X}$, f, $\hat{f}$, $\mathbf{e}$ as well as the set $\mathbf{E}$ of bound constraints over $\mathbf{e}$. For a given machine ε, one has $\mathbf{E} := [-\varepsilon, \varepsilon]^m$, with m being the number of error variables. This algorithm actually relies on the CSSOS hierarchy from Chapter 3, thus `roundoff_bound` also takes as input a relaxation order $r \in \mathbb{N}^*$. The algorithm provides as output an interval enclosure of the error $\hat{f}(\mathbf{x}, \mathbf{e}) - f(\mathbf{x})$ over $\mathbf{K} := \mathbf{X} \times \mathbf{E}$. From this interval $[f_{\min}^r, f_{\max}^r]$, one can compute $|f|_{\max}^r := \max\{-f_{\min}^r, f_{\max}^r\}$, which is a sound upper bound of the maximal absolute error $|\Delta|_{\max} := \max_{(\mathbf{x}, \mathbf{e}) \in \mathbf{K}} |\hat{f}(\mathbf{x}, \mathbf{e}) - f(\mathbf{x})|$.

Algorithm 2 `roundoff_bound`

Require: input variables $\mathbf{x}$, input constraints $\mathbf{X}$, nonlinear expression f, rounded expression $\hat{f}$, error variables $\mathbf{e}$, error constraints $\mathbf{E}$, relaxation order r

Ensure: interval enclosure of the error $\hat{f} - f$ over $\mathbf{K} := \mathbf{X} \times \mathbf{E}$

  1: Define the absolute error $\Delta(\mathbf{x}, \mathbf{e}) := \hat{f}(\mathbf{x}, \mathbf{e}) - f(\mathbf{x})$

  2: Compute $\ell(\mathbf{x}, \mathbf{e}) := \Delta(\mathbf{x}, 0) + \sum_{j=1}^m \frac{\partial \Delta(\mathbf{x}, \mathbf{e})}{\partial e_j}(\mathbf{x}, 0) \, e_j$

  3: Define $h := \Delta - \ell$

  4: $[\underline{h}, \overline{h}] := \mathtt{ia\_bound}(h, \mathbf{K})$ $\triangleright$ Compute bounds for h

  5: $[\ell_{\min}^r, \ell_{\max}^r] := \mathtt{cs\_sdp}(\ell, \mathbf{K}, r)$ $\triangleright$ Compute bounds for ℓ

  6: **return** $[\ell_{\min}^r + \underline{h}, \ell_{\max}^r + \overline{h}]$

After defining the absolute roundoff error $\Delta := \hat{f} - f$ (Step 1), one decomposes Δ as the sum of an expression ℓ which is affine with respect to the error variable $\mathbf{e}$ and a remainder h. One way to obtain ℓ is to compute the vector of partial derivatives of Δ with respect to $\mathbf{e}$ evaluated at $(\mathbf{x}, 0)$ and finally to take the inner product of this vector and $\mathbf{e}$ (Step 2). Then, the idea is to compute a precise bound of ℓ and a coarse bound of h. The underlying reason is that h involves error term products of degree greater than 2 (e.g. $e_1 e_2$), yielding an interval enclosure of *a priori* much smaller width, compared to the interval enclosure of ℓ. One obtains the interval enclosure of h using the procedure `ia_bound` implementing basic interval arithmetic (Step 4) to bound the remainder of the multivariate Taylor expansion of Δ with respect to $\mathbf{e}$, expressed as a combination of the second-error derivatives (similar as in [SJRG15]). The algorithm roundoff_bound is very similar to the algorithm of FPTaylor [SJRG15], except that SDP based techniques are used instead of the global optimization procedure from [SJRG15]. Note that overflow and denormal are neglected here but one could handle them, as in [SJRG15], by adding additional error variables and discarding the related terms using naive interval arithmetic.

The bounds of ℓ are provided through the cs_sdp procedure, which solves two instances of (3.7), at relaxation order r. We now give more explanation about this procedure. We can map each input variable x_i to the integer i, for all $i \in [n]$, as well as each error variable e_j to $n+j$, for all $j \in [m]$. Then, define the sets $I_1 := [n] \cup \{n+1\}, \ldots, I_m := [n] \cup \{n+m\}$. Here, we take advantage of the csp of ℓ by using m distinct sets of cardinality $n+1$ rather than a single one of cardinality $n+m$, i.e., the total number of variables. Note that these subsets satisfy (1.8) and one can write $\ell(\mathbf{x}, \mathbf{e}) = \Delta(\mathbf{x}, \mathbf{0}) + \sum_{j=1}^{m} \frac{\partial \Delta(\mathbf{x}, \mathbf{e})}{\partial e_j}(\mathbf{x}, \mathbf{0}) e_j$. After noticing that $\Delta(\mathbf{x}, \mathbf{0}) = \hat{f}(\mathbf{x}, \mathbf{0}) - f(\mathbf{x}) = 0$, one can scale the optimization problems by writing

$$\ell(\mathbf{x}, \mathbf{e}) = \sum_{j=1}^{m} s_j(\mathbf{x}) e_j = \varepsilon \sum_{j=1}^{m} s_j(\mathbf{x}) \frac{e_j}{\varepsilon}, \tag{4.2}$$

with $s_j(\mathbf{x}) := \frac{\partial \Delta(\mathbf{x}, \mathbf{e})}{\partial e_j}(\mathbf{x}, \mathbf{0})$, for all $j \in [m]$. Replacing $\mathbf{e}$ by $\mathbf{e}/\varepsilon$ leads to computing an interval enclosure of ℓ/ε over $\mathbf{K}' := \mathbf{X} \times [-1, 1]^m$. As usual from Assumption 2.2, there exists an integer $N > 0$ such that $N - \sum_{i=1}^{n} x_i^2 \geq 0$, as the input variables satisfy box constraints. Moreover, to fulfill Assumption 3.1, one encodes $\mathbf{K}'$ as follows:

$$\mathbf{K}' := \{(\mathbf{x}, \mathbf{e}) \in \mathbb{R}^{n+m} : g_1(\mathbf{x}) \geq 0, \ldots, g_l(\mathbf{x}) \geq 0,$$
$$g_{l+1}(\mathbf{x}, e_1) \geq 0, \ldots, g_{l+m}(\mathbf{x}, e_m) \geq 0\},$$

with $g_{l+j}(\mathbf{x}, e_j) := N + 1 - \sum_{i=1}^{n} x_i^2 - e_j^2$, for all $j \in [m]$. The index set of variables involved in g_j is $[n]$ for all $j \in [l]$. The index set of variables involved in g_{l+j} is I_j for all $j \in [m]$.

Then, one can compute a lower bound of the minimum of $\ell'(\mathbf{x}, \mathbf{e}) := \ell(\mathbf{x}, \mathbf{e})/\varepsilon = \sum_{j=1}^{m} s_j(\mathbf{x}) e_j$ over $\mathbf{K}'$ by solving the following CSSOS problem:

$$\begin{aligned}
\ell_{\min}'^r := \quad &\sup_{b, \sigma_j} \quad b \\
&\text{s.t.} \quad \ell' - b = \sigma_0 + \sum_{j=1}^{l+m} \sigma_j g_j \\
&\qquad \sigma_0 \in \sum_{j=1}^{m} \Sigma[(\mathbf{x}, \mathbf{e}), I_j] \\
&\qquad \sigma_j \in \Sigma[(\mathbf{x}, \mathbf{e}), J_j], \quad j \in [l+m] \\
&\qquad \deg(\sigma_j g_j) \leq 2r, \quad j = 0, \ldots, l+m.
\end{aligned} \tag{4.3}$$

A feasible solution of Problem (4.3) ensures the existence of $\sigma^1 \in \Sigma[(\mathbf{x}, e_1)]$, $\ldots, \sigma^m \in \Sigma[(\mathbf{x}, e_m)]$ such that $\sigma_0 = \sum_{j=0}^{m} \sigma^j$, allowing the following reformulation:

$$
\begin{aligned}
\ell_{\min}^{\prime r} = \quad &\sup_{b, \sigma^j, \sigma_j} \quad b \\
&\text{s.t.} \quad \ell' - b = \sum_{j=1}^{m} \sigma^j + \sum_{j=1}^{l+m} \sigma_j g_j \\
&\qquad \sigma_j \in \Sigma[\mathbf{x}], \quad j \in [m] \\
&\qquad \sigma^j \in \Sigma[(\mathbf{x}, e_j)], \deg(\sigma^j) \le 2r, \quad j \in [m] \\
&\qquad \deg(\sigma_j g_j) \le 2r, \quad j \in [l + m].
\end{aligned}
\tag{4.4}
$$

An upper bound $\ell_{\max}^{\prime r}$ can be obtained by replacing sup with inf and $\ell' - b$ by $b - \ell'$ in Problem (4.4). Our optimization procedure cs_sdp computes the lower bound $\ell_{\min}^{\prime r}$ as well as an upper bound $\ell_{\max}^{\prime r}$ of ℓ' over $\mathbf{K}'$, and then returns the interval $[\varepsilon\, \ell_{\min}^{\prime r}, \varepsilon\, \ell_{\max}^{\prime r}]$, which is a sound enclosure of the values of ℓ over $\mathbf{K}$.

We emphasize two advantages of the decomposition $\Delta = \ell + h$ and more precisely of the linear dependency of ℓ with respect to $\mathbf{e}$: scalability and robustness to SDP numerical issues. First, no computation is required to determine the csp of ℓ, by comparison to the general case. Thus, it becomes much easier to handle the optimization of ℓ with the sparse SDP (4.4) rather than with the corresponding instance of the dense relaxation $(\mathbf{P}^r)$, given in (2.7). While the latter involves $\binom{n+m+2r}{2r}$ SDP variables, the former involves only $m \binom{n+1+2r}{2r}$ SDP variables, ensuring the scalability of our framework. In addition, the linear dependency of ℓ with respect to $\mathbf{e}$ allows us to scale the error variables and optimize over a set of variables lying in $\mathbf{K}' := \mathbf{X} \times [-1, 1]^m$. It ensures that the range of input variables does not significantly differ from the range of error variables. This condition is mandatory in considering SDP relaxations because most SDP solvers (e.g., MOSEK [ART03]) are implemented using double precision floating-point. It is impossible to optimize ℓ over $\mathbf{K}$ (rather than ℓ' over $\mathbf{K}'$) when the maximal value ε of error variables is less than 2^{-53}, due to the fact that SDP solvers would treat each error variable term as 0, and consequently ℓ as the zero polynomial. Thus, this decomposition insures our framework against numerical issues related to finite-precision implementation of SDP solvers.

Let us consider the interval $[\ell_{\min}, \ell_{\max}]$, with $\ell_{\min} := \inf_{(\mathbf{x}, \mathbf{e}) \in \mathbf{K}} \ell(\mathbf{x}, \mathbf{e})$ and $\ell_{\max} := \sup_{(\mathbf{x}, \mathbf{e}) \in \mathbf{K}} \ell(\mathbf{x}, \mathbf{e})$. The next lemma states that one can approximate this interval as closely as desired using the cs_sdp procedure.

Lemma 4.1 (Convergence of the cs_sdp procedure) *Let $[\ell_{\min}^r, \ell_{\max}^r]$ be the interval enclosure returned by the procedure* cs_sdp$(\ell, \mathbf{K}, r)$. *Then the sequence $([\ell_{\min}^r, \ell_{\max}^r])_{r \in \mathbb{N}}$ converges to $[\ell_{\min}, \ell_{\max}]$ when r goes to infinity.*

The proof of Lemma 4.1 is based on the fact that the assumptions of Theorem 3.8 are fulfilled for our specific roundoff error problem. This result guarantees asymptotic convergence to the exact enclosure of ℓ when the relaxation order r tends to infinity. However, it is more reasonable in practice to keep this order as small as possible to obtain tractable SDP relaxations. Hence, we generically solve each instance of Problem (4.4) at the minimal relaxation order, that is $r_{min} = \max\{\lceil \deg \ell/2 \rceil, \lceil \deg(g_j)/2 \rceil, j = 1, \ldots, l+m\}$. Afterwards, we can rely on the Coq computer assistant to obtain formally certified upper bounds for the roundoff error; see [MCD17, §2.3] for more details.

4.3   Overview of numerical experiments

We present an overview of our method and of the capabilities of related techniques, using an example. Consider a program implementing the following polynomial expression f:

$$f(\mathbf{x}) := x_2 \times x_5 + x_3 \times x_6 - x_2 \times x_3 - x_5 \times x_6$$
$$+ x_1 \times (-x_1 + x_2 + x_3 - x_4 + x_5 + x_6),$$

where the six-variable vector $\mathbf{x} := (x_1, x_2, x_3, x_4, x_5, x_6)$ is the input of the program. Here the set $\mathbf{X}$ of possible input values is a product of closed intervals: $\mathbf{X} = [4.00, 6.36]^6$. This function f together with the set $\mathbf{X}$ appear in many inequalities arising from the proof of the Kepler Conjecture [Hal06], yielding challenging global optimization problems.

The polynomial expression f is obtained by performing 15 basic operations (1 negation, 3 subtractions, 6 additions and 5 multiplications). When executing this program with a set of floating-point numbers $\hat{\mathbf{x}} := (\hat{x}_1, \hat{x}_2, \hat{x}_3, \hat{x}_4, \hat{x}_5, \hat{x}_6) \in \mathbf{X}$, one actually computes a floating-point result $\hat{f}$, where all operations $+, -, \times$ are replaced by the respectively associated floating-point operations $\oplus, \ominus, \otimes$. The results of these operations comply with IEEE 754 standard arithmetic [IEE08]. Here, for the sake of clarity, we do not consider real input variables. For instance, (in the absence of underflow) one can write $\hat{x}_2 \otimes \hat{x}_5 = (x_2 \times x_5)(1 + e_1)$, by introducing an error variable e_1 such that $-\varepsilon \leq e_1 \leq \varepsilon$, where the bound ε is the machine precision (e.g., $\varepsilon = 2^{-24}$ for single precision). One would like to bound the absolute roundoff error $|\Delta(\mathbf{x}, \mathbf{e})| := |\hat{f}(\mathbf{x}, \mathbf{e}) - f(\mathbf{x})|$ over all possible input variables $\mathbf{x} \in \mathbf{X}$ and error variable $e_1, \ldots, e_{15} \in [-\varepsilon, \varepsilon]$. Let us define $\mathbf{E} := [-\varepsilon, \varepsilon]^{15}$ and $\mathbf{K} := \mathbf{X} \times \mathbf{E}$. Then our bound problem can be cast as finding the maximum $|\Delta|_{max}$ of $|\Delta|$ over $\mathbf{K}$, yielding the following nonlinear optimization problem:

$$|\Delta|_{max} := \max_{(\mathbf{x,e}) \in \mathbf{K}} |\Delta(\mathbf{x,e})|$$
$$= \max\{-\min_{(\mathbf{x,e}) \in \mathbf{K}} \Delta(\mathbf{x,e}), \max_{(\mathbf{x,e}) \in \mathbf{K}} \Delta(\mathbf{x,e})\}. \qquad (4.5)$$

One can directly try to solve these two POPs using classical SDP relaxations [Las01]. As in [SJRG15], one can also decompose the error term Δ as the sum of a term $\ell(\mathbf{x,e})$, which is affine with respect to $\mathbf{e}$, and a nonlinear term $h(\mathbf{x,e}) := \Delta(\mathbf{x,e}) - \ell(\mathbf{x,e})$. Then the triangular inequality yields:

$$|\Delta|_{max} \le \max_{(\mathbf{x,e}) \in \mathbf{K}} |\ell(\mathbf{x,e})| + \max_{(\mathbf{x,e}) \in \mathbf{K}} |h(\mathbf{x,e})|. \qquad (4.6)$$

It follows for this example that

$$\ell(\mathbf{x,e}) = x_2 x_5 e_1 + x_3 x_6 e_2 + (x_2 x_5 + x_3 x_6)e_3 + \cdots + f(\mathbf{x})e_{15} = \sum_{i=1}^{15} s_i(\mathbf{x})e_i,$$

with $s_1(\mathbf{x}) := x_2 x_5, s_2(\mathbf{x}) := x_3 x_6, \ldots, s_{15}(\mathbf{x}) := f(\mathbf{x})$. The *Symbolic Taylor Expansions* method [SJRG15] consists of using a simple branch and bound algorithm based on interval arithmetic to compute a rigorous interval enclosure of each polynomial s_i, $i \in [15]$, over $\mathbf{X}$ and finally obtain an upper bound of $|\ell| + |h|$ over $\mathbf{K}$. In contrast, our method uses sparse SDP relaxations for polynomial optimization (derived from [Las06]) to bound $|\ell|$ and basic interval arithmetic as in [SJRG15] to bound $|h|$ (i.e., we use interval arithmetic to bound second-order error terms in the multivariate Taylor expansion of Δ with respect to $\mathbf{e}$).

- A direct attempt to solve the two polynomial problems occurring in Equation (4.5) fails as the SDP solver (in our case SDPA [YFN+10]) runs out of memory.
- Using our method implemented in the Real2Float tool,[1] one obtains an upper bound of 760ε for $|\ell| + |h|$ over $\mathbf{K}$ in 0.15 seconds. This bound is provided together with a certificate which can be formally checked inside the Coq proof assistant in 0.20 seconds.
- After normalizing the polynomial expression and using basic interval arithmetic, one obtains 8 times more quickly a coarser bound of 922ε.
- Symbolic Taylor expansions implemented in FPTaylor [SJRG15] provide a more precise bound of 721ε, but the analysis time is 28 times slower than with our implementation. Formal verification of this bound inside the $\mathrm{Hol} - \mathrm{light}$ proof assistant takes 28 seconds, which is 139 times slower than proof checking with Real2Float inside Coq.

---

[1]https://forge.ocamlcore.org/projects/nl-certify/.

One can obtain an even more precise bound of 528ε (but 37 times slower than with our implementation) by turning on the improved rounding model of FPTaylor and limiting the number of branch and bound iterations to 10000. The drawback of this bound is that it cannot be formally verified.

- Finally, a slightly coarser bound of 762ε is obtained with the Rosa real compiler [DK14], but the analysis is 19 times slower than with our implementation and we cannot get formal verification of this bound.

4.4   Notes and sources

To obtain lower bounds on roundoff errors, one can rely on testing approaches, such as meta-heuristic search [BdA+12] or under-approximation tools (e.g., s3fp [CGRS14]). Here, we are interested in efficiently handling the complementary over-approximation problem, namely to obtain precise upper bounds on the error. This problem boils down to finding tight abstractions of linearities or nonlinearities while being able to bound the resulting approximations in an efficient way. For computer programs consisting of linear operations, automatic error analysis can be obtained with well-studied optimization techniques based on SAT/SMT solvers [HGBK12] and affine arithmetic [DGP+09]. However, nonlinear operations are key to many interesting computational problems arising in physics, biology, controller implementations and global optimization. Two promising frameworks have been designed to provide upper bounds for roundoff errors of nonlinear programs. The corresponding algorithms rely on Taylor-interval methods [SJRG15], implemented in the FPTaylor tool, and on combining SMT with interval arithmetic [DK14], implemented in the Rosa real compiler. We refer the interested reader to [MCD17, §4] for more details on the extensive experimental evaluation that we performed.

Bibliography

[ART03]   Erling D Andersen, Cornelis Roos, and Tamas Terlaky. On implementing a primal-dual interior-point method for conic quadratic optimization. *Mathematical Programming*, 95(2):249–277, 2003.

[BdA+12]   Mateus Borges, Marcelo d'Amorim, Saswat Anand, David Bushnell, and Corina S. Pasareanu. Symbolic execution with interval solving and meta-heuristic search. In *Proceedings of the 2012 IEEE Fifth International Conference on Software Testing, Verification and Validation*, ICST '12, pages 111–120, Washington, DC, USA, 2012. IEEE Computer Society.

[CGRS14]   Wei-Fan Chiang, Ganesh Gopalakrishnan, Zvonimir Rakamaric, and Alexey Solovyev. Efficient search for inputs causing high floating-point errors. In *Proceedings of the 19th ACM SIGPLAN Symposium on Principles and Practice of Parallel Programming*, PPoPP '14, pages 43–52, New York, NY, USA, 2014. ACM.

[DGP+09]   David Delmas, Eric Goubault, Sylvie Putot, Jean Souyris, Karim Tekkal, and Franck Védrine. Towards an industrial use of fluctuat on safety-critical avionics software. In María Alpuente, Byron Cook, and Christophe Joubert, editors, *Formal Methods for Industrial Critical Systems*, volume 5825 of *Lecture Notes in Computer Science*, pages 53–69. Springer Berlin Heidelberg, 2009.

[DK14]   Eva Darulova and Viktor Kuncak. Sound Compilation of Reals. In *Proceedings of the 41st ACM SIGPLAN-SIGACT Symposium on Principles of Programming Languages*, POPL '14, pages 235–248, New York, NY, USA, 2014. ACM.

[Hal06]   Thomas C. Hales. Introduction to the flyspeck project. In Thierry Coquand, Henri Lombardi, and Marie-Françoise Roy, editors, *Mathematics, Algorithms, Proofs*, number 05021 in Dagstuhl Seminar Proceedings, Dagstuhl, Germany, 2006.

[HGBK12] Leopold Haller, Alberto Griggio, Martin Brain, and Daniel Kroening. Deciding floating-point logic with systematic abstraction. In *Formal Methods in Computer-Aided Design (FMCAD)*, pages 131–140, 2012.

[Hig02] N.J. Higham. *Accuracy and Stability of Numerical Algorithms: Second Edition*. Society for Industrial and Applied Mathematics, 2002.

[IEE08] IEEE. IEEE Standard for Floating-Point Arithmetic. *IEEE Std 754-2008*, pages 1–70, 2008.

[Las01] Jean-Bernard Lasserre. Global Optimization with Polynomials and the Problem of Moments. *SIAM Journal on Optimization*, 11(3):796–817, 2001.

[Las06] Jean B Lasserre. Convergent sdp-relaxations in polynomial optimization with sparsity. *SIAM Journal on Optimization*, 17(3):822–843, 2006.

[MAGW15] Victor Magron, Xavier Allamigeon, Stéphane Gaubert, and Benjamin Werner. Certification of real inequalities: templates and sums of squares. *Mathematical Programming*, 151(2):477–506, 2015.

[MCD17] Victor Magron, George Constantinides, and Alastair Donaldson. Certified roundoff error bounds using semidefinite programming. *ACM Trans. Math. Software*, 43(4):Art. 34, 31, 2017.

[SJRG15] Alexey Solovyev, Charles Jacobsen, Zvonimir Rakamaric, and Ganesh Gopalakrishnan. Rigorous Estimation of Floating-Point Round-off Errors with Symbolic Taylor Expansions. In Nikolaj Bjorner and Frank de Boer, editors, *Proceedings of the 20th International Symposium on Formal Methods (FM)*, volume 9109 of *Lecture Notes in Computer Science*, pages 532–550. Springer, 2015.

[YFN+10] M. Yamashita, K. Fujisawa, K. Nakata, M. Nakata, M. Fukuda, K. Kobayashi, and K. Goto. A high-performance software package for semidefinite programs : SDPA7. Technical report, Dept. of Information Sciences, Tokyo Inst. Tech., 2010.

Chapter 5

Application in deep networks

The Lipschitz constant of a network plays an important role in many applications of deep learning, such as robustness certification and Wasserstein Generative Adversarial Network. We introduce an SDP hierarchy to estimate the global and local Lipschitz constant of a multiple layer deep neural network. The novelty is to combine a polynomial lifting for ReLU function derivatives with a weak generalization of Putinar's positivity certificate. We empirically demonstrate that our method provides a trade-off with respect to the state-of-the-art LP approach, and in some cases we obtain better bounds in less time.

5.1 Multiple layer networks

We focus on multiple layer networks with ReLU activations. Recall that a function f, defined on a convex set $\mathcal{X} \subseteq \mathbb{R}^n$, is L-Lipschitz with respect to the norm $\| \cdot \|$ if for all $\mathbf{x}, \mathbf{z} \in \mathcal{X}$, we have $|f(\mathbf{x}) - f(\mathbf{z})| \leq L\|\mathbf{x} - \mathbf{z}\|$. The Lipschitz constant of f with respect to norm $\| \cdot \|$, denoted by $L_f^{\|\cdot\|}$, is the infimum of all those valid L's:

$$L_f^{\|\cdot\|} := \inf \{L : \forall \mathbf{x}, \mathbf{z} \in \mathcal{X}, |f(\mathbf{x}) - f(\mathbf{z})| \leq L\|\mathbf{x} - \mathbf{z}\|\}. \qquad (5.1)$$

We denote by F the multiple layer neural network, m the number of hidden layers, $p_0, p_1, \ldots, p_m$ the number of nodes in the input layer and each hidden layer. For simplicity, $(p_0, p_1, \ldots, p_m)$ will denote the layer structure of network F. Let $\mathbf{x}_0$ be the initial input, and $\mathbf{x}_1, \ldots, \mathbf{x}_m$ be the activation vectors in each hidden layer. Each $\mathbf{x}_i$, $i \in [m]$, is obtained by a weight $\mathbf{A}_i$, a bias $\mathbf{b}_i$, and an activation function a, i.e., $\mathbf{x}_i = a(\mathbf{z}_i)$ with $\mathbf{z}_i = \mathbf{A}_i \mathbf{x}_{i-1} + \mathbf{b}_i$; see Figure 5.1.

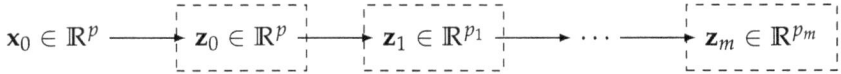

Figure 5.1: Description of a multiple layer neural network.

We only consider coordinatewise application of the ReLU activation function, defined as $\text{ReLU}(x) = \max\{0, x\}$ for $x \in \mathbb{R}$. The ReLU function is non-smooth, and we define its generalized derivative as the set-valued function $G(x)$ such that $G(x) = 1$ for $x > 0$, $G(x) = 0$ for $x < 0$ and $G(x) = \{0, 1\}$ for $x = 0$. The key reason why neural networks with ReLU activation function can be tackled using polynomial optimization techniques is semialgebraicity of the ReLU function, i.e., it can be expressed with a system of polynomial (in)equalities. For $x, u \in \mathbb{R}$, we have $u = \text{ReLU}(x) = \max\{0, x\}$ if and only if $u(u - x) = 0, u \geq x, u \geq 0$. Similarly, one can exploit the semialgebraicity of its derivative ReLU'; see Figure 5.2 and Figure 5.3.

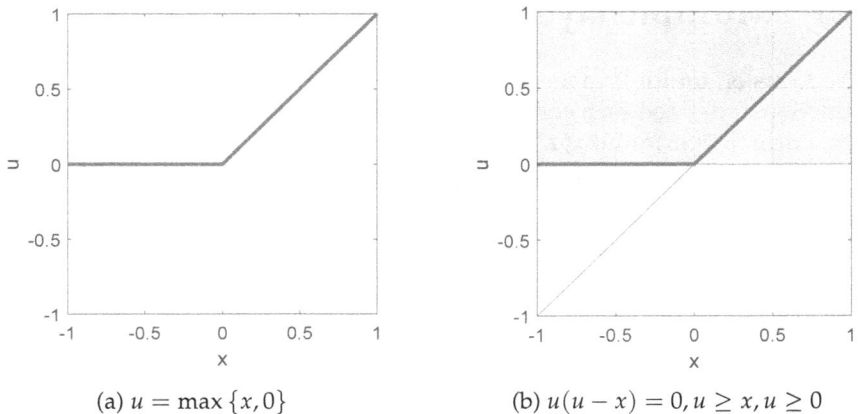

(a) $u = \max\{x, 0\}$ (b) $u(u - x) = 0, u \geq x, u \geq 0$

Figure 5.2: ReLU (left) and its semialgebraicity (right).

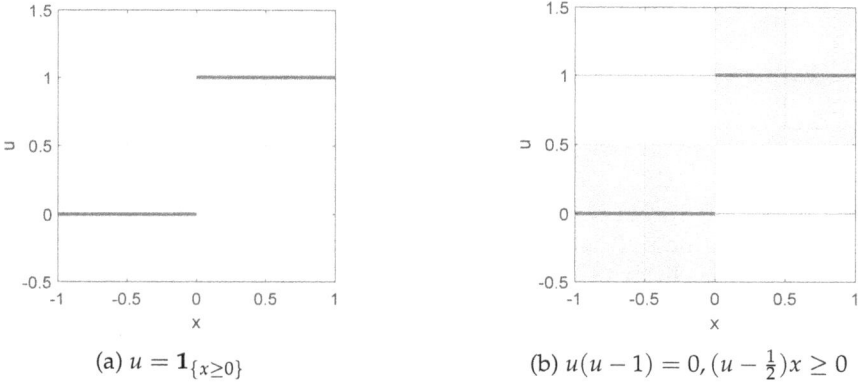

(a) $u = \mathbf{1}_{\{x \geq 0\}}$ (b) $u(u-1) = 0, (u - \frac{1}{2})x \geq 0$

Figure 5.3: ReLU$'$ (left) and its semialgebraicity (right).

We assume that the last layer in our neural network is a softmax layer with K entries, that is, the network is a classifier for K labels. For each label $k \in \{1, \ldots, K\}$, the score of label k is obtained by an affine product with the last activation vector, i.e., $c_k^\mathsf{T} x_m$ for some $c_k \in \mathbb{R}^{p_m}$. The final output is the label with the highest score, i.e., $z = \arg\max_k c_k^\mathsf{T} x_m$. The product xz of two vectors x and z is considered as the coordinate-wise product.

5.2   Lipschitz constants

Suppose we train a neural network F for K-classifications and denote by A_i, b_i, c_k its parameters. Thus for an input $x_0 \in \mathbb{R}^{p_0}$, the targeted score of label k can be expressed as $F_k(x_0) = c_k^\mathsf{T} x_m$, where $x_i = \text{ReLU}(A_i x_{i-1} + b_i)$, for $i \in [m]$. Let $z_i = A_i x_{i-1} + b_i$ for $i \in [m]$. By applying the chain rule on the non-smooth function F_k, we obtain a set valued map for F_k at point any x_0 as $G_{F_k}(x_0) = (\prod_{i=1}^m A_i^\mathsf{T} \text{Diag}(G(z_i))) c_k$.

We fix a targeted label (label 1 for example) and omit the symbol k for simplicity. We define $L_F^{\|\cdot\|}$ as the supremum of the gradient's dual norm:

$$L_F^{\|\cdot\|} = \sup_{x_0 \in \Omega, \, v \in G_{F_k}(x_0)} \|v\|_* = \sup_{x_0 \in \Omega} \left\| \left(\prod_{i=1}^m A_i^\mathsf{T} \text{Diag}(G(z_i)) \right) c \right\|_*, \quad (5.2)$$

where Ω is the convex input space, and $\|\cdot\|_*$ is the dual norm of $\|\cdot\|$, which is defined by $\|x\|_* := \sup_{\|t\| \leq 1} |\langle t, x \rangle|$ for all $x \in \mathbb{R}^n$. In general, the chain rule cannot be applied to composition of non-smooth functions [KL18, BP21]. Hence the formulation of G_{F_k} and (5.2) may lead to incorrect gradients and bounds on the Lipschitz constant of the networks. The following ensures that this is not the case and that the approach is sound.

> **Theorem 5.1** *If Ω is convex, then $L_F^{\|\cdot\|}$ is a Lipschitz constant for F_k on Ω.*

When $\Omega = \mathbb{R}^n$, $L_F^{\|\cdot\|}$ is the *global* Lipschitz constant of F with respect to norm $\|\cdot\|$. In many cases we are also interested in the *local* Lipschitz constant of a neural network constrained in a small neighborhood of a fixed input $\bar{x}_0$. In this situation the input space Ω is often the ball around $\bar{x}_0 \in \mathbb{R}^n$ with radius ε: $\Omega = \{x : \|x - \bar{x}_0\| \leq \varepsilon\}$. In particular, with the L_∞-norm (and using $l \leq x \leq u \Leftrightarrow (x - l)(x - u) \leq 0$), the input space Ω is the basic closed semialgebraic set:

$$\Omega = \{x \in \mathbb{R}^n : (x - \bar{x}_0 + \varepsilon)(x - \bar{x}_0 - \varepsilon) \leq 0\}. \tag{5.3}$$

In view of Theorem 5.1 and (5.2), the *Lipschitz constant estimation problem (LCEP)* for neural networks with respect to the norm $\|\cdot\|$ is the following POP:

$$\begin{cases} \sup_{x_i, u_i, t} & t^\top \left(\prod_{i=1}^m A_i^\top \operatorname{Diag}(u_i) \right) c \\ \text{s.t.} & u_i(u_i - 1) = 0, (u_i - 1/2)(A_i x_{i-1} + b_i) \geq 0, i \in [m] \\ & x_{i-1}(x_{i-1} - A_{i-1}x_{i-2} - b_{i-1}) = 0 \\ & x_{i-1} \geq 0, x_{i-1} \geq A_{i-1}x_{i-2} + b_{i-1}, i = 2, \dots, m \\ & t^2 \leq 1, (x_0 - \bar{x}_0 + \varepsilon)(x_0 - \bar{x}_0 - \varepsilon) \leq 0. \end{cases} \tag{5.4}$$

In [LRC20] the authors only use the constraint $0 \leq u_i \leq 1$ on the variables u_i, only capturing the Lipschitz character of the considered activation function. We could use the same constraints, and this would allow to use activations which do not have semi-algebraic representations such as the Exponential Linear Unit (ELU). However, such a relaxation, despite very general, is a lot coarser than the one we propose. Indeed, (5.4) treats an *exact formulation* of the generalized derivative of the ReLU function by exploiting its semialgebraic character.

5.3 Nearly sparse problems

For illustration purpose, consider 1-hidden layer networks. Then in (5.4) we can define natural subsets $I_i = \{u_1^{(i)}, x_0\}$, $i \in [p_1]$ (with respect to constraints $u_1(u_1 - 1) = 0$, $(u_1 - 1/2)(A_1 x_0 + b_1) \geq 0$, and $(x_0 - \bar{x}_0 + \varepsilon)(x_0 - \bar{x}_0 - \varepsilon) \leq 0$); and $J_j = \{t^{(j)}\}$, $j \in [p_0]$ (with respect to constraints $t^2 \leq 1$). Clearly, I_i, J_j satisfy the RIP condition (1.8) and are subsets with smallest possible size. Recall that $x_0 \in \mathbb{R}^{p_0}$. Hence $|I_i| = 1 + p_0$ and the maximum size of the PSD matrices involved in the sparse Lasserre's hierarchy $\mathbf{P}_{cs}^r$, given in (3.6), is $\binom{1+p_0+r}{r}$. Therefore, as in real deep neural networks p_0 can be as large as 1000, the second-order sparse Lasserre's hierarchy $\mathbf{P}_{cs}^2$, cannot be implemented in practice.

In fact (5.4) can be considered as a "nearly sparse" POP, i.e., a sparse POP with some additional "bad" constraints that violate the sparsity assumptions. More precisely, suppose that f, g_i and subsets I_k satisfy Assumption 3.1. Let g be a polynomial that violates Assumption 3.1(iii), i.e., RIP (1.8). Then we call the POP

$$\inf_{\mathbf{x} \in \mathbb{R}^n} \{f(\mathbf{x}) : g(\mathbf{x}) \geq 0, g_i(\mathbf{x}) \geq 0, i \in [m]\}, \qquad \text{(Nearly)}$$

a *nearly sparse* POP because only one constraint, namely $g \geq 0$, does not satisfy the sparsity pattern corresponding to the RIP (1.8). This single "bad" constraint $g \geq 0$ precludes us from applying the sparse Lasserre hierarchy (3.6).

In this situation, we propose a heuristic method which can be applied to problems with arbitrary many constraints that possibly destroy the sparsity pattern. The key idea of our algorithm is: **(i)** Keep the "nice" sparsity pattern defined without the bad constraints; **(ii)** Associate only low-order localizing matrix constraints to the "bad" constraints. In brief, the r-th order *heuristic hierarchy (HR-r)* reads as

$$\begin{cases} \inf_{\mathbf{y}} & L_{\mathbf{y}}(f) \\ \text{s.t.} & \mathbf{M}_1(\mathbf{y}) \succeq 0 \\ & \mathbf{M}_r(\mathbf{y}, I_k) \succeq 0, \quad k \in [l] \\ & \mathbf{M}_{r-d_i}(g_i \, \mathbf{y}, I_{k(i)}) \succeq 0, \quad i \in [m] \\ & L_{\mathbf{y}}(g) \geq 0, \quad y_0 = 1. \end{cases} \qquad \text{(HR-r)}$$

We already have a sparsity pattern with subsets I_k and an additional "bad" constraint $g \geq 0$ (assumed to be quadratic). Then we consider the sparse moment relaxations (3.6) applied to (Nearly) *without* the bad constraint $g \geq 0$ and simply add two constraints: (i) the moment constraint $\mathbf{M}_1(\mathbf{y}) \succeq 0$ (with full dense first-order moment matrix $\mathbf{M}_1(\mathbf{y})$), and (ii) the linear moment inequality constraint $L_{\mathbf{y}}(g) \geq 0$ (which is the lowest-order localizing matrix constraint $\mathbf{M}_0(g \, \mathbf{y}) \succeq 0$).

To see why the full moment constraint $\mathbf{M}_1(\mathbf{y}) \succeq 0$ is needed, consider the following toy problem:

$$\inf_{\mathbf{x} \in \mathbb{R}^3} \{x_1 x_2 + x_2 x_3 : x_1^2 + x_2^2 \leq 1, x_2^2 + x_3^2 \leq 1\}. \qquad (5.5)$$

Define the subsets $I_1 = \{1,2\}$, $I_2 = \{2,3\}$. It is easy to check that Assumption 3.1 holds. Define

$$\mathbf{y} = \{y_{000}, y_{100}, y_{010}, y_{001}, y_{200}, y_{110}, y_{101}, y_{020}, y_{011}, y_{002}\} \in \mathbb{R}^{10}.$$

For $r = 1$, the first-order dense moment matrix reads as

$$\mathbf{M}_1(\mathbf{y}) = \begin{bmatrix} y_{000} & y_{100} & y_{010} & y_{001} \\ y_{100} & y_{200} & y_{110} & y_{101} \\ y_{010} & y_{110} & y_{020} & y_{011} \\ y_{001} & y_{101} & y_{011} & y_{002} \end{bmatrix},$$

whereas the sparse moment matrix $\mathbf{M}_1(\mathbf{y}, I_1)$ (resp. $\mathbf{M}_1(\mathbf{y}, I_2)$) is the sub-matrix of $\mathbf{M}_1(\mathbf{y})$ taking red and pink (resp. blue and pink) entries. That is, $\mathbf{M}_1(\mathbf{y}, I_1)$ and $\mathbf{M}_1(\mathbf{y}, I_2)$ are submatrices of $\mathbf{M}_1(\mathbf{y})$, obtained by restricting to rows and columns concerned with subsets I_1 and I_2 only.

Now suppose that we need to consider an additional "bad" constraint $(1 - x_1 - x_2 - x_3)^2 = 0$. After developing $L_{\mathbf{y}}(g)$, one needs to consider the moment variable y_{101} corresponding to the monomial $x_1 x_3$ in the expansion of $g = (1 - x_1 - x_2 - x_3)^2$, and y_{101} does *not* appear in the moment matrices $\mathbf{M}_1(\mathbf{y}, I_1)$ and $\mathbf{M}_1(\mathbf{y}, I_2)$ because x_1 and x_3 are not in the same subset. However y_{101} appears in $\mathbf{M}_1(\mathbf{y})$, which is of size $n + 1$.

Now let us see how this works for problem (5.4). First introduce new variables $\mathbf{z}_i$ with associated constraints $\mathbf{z}_i - \mathbf{A}_i \mathbf{x}_{i-1} - \mathbf{b}_i = 0$, so that all "bad" constraints are affine. Equivalently, we may and will consider the single "bad" constraint $g \geq 0$ with

$$g(\mathbf{z}_1, \ldots, \mathbf{x}_0, \mathbf{x}_1, \ldots) = -\sum_i \|\mathbf{z}_i - \mathbf{A}\mathbf{x}_{i-1} - \mathbf{b}_i\|^2$$

and solve (HR-r). We briefly sketch the rationale behind this reformulation. Let $(\mathbf{y}^r)_{r \in \mathbb{N}}$ be a sequence of optimal solutions of (HR-r). If $r \to \infty$, then $\mathbf{y}^r \to \mathbf{y}$ (possibly for a subsequence $(r_k)_{k \in \mathbb{N}}$), and $\mathbf{y}$ corresponds to the moment sequence of a measure μ, supported on $\{(\mathbf{x}, \mathbf{z}) : g_i(\mathbf{x}, \mathbf{z}) \geq 0, i \in [p]; \int g \, d\mu \geq 0\}$. But as $-g$ is a square, $\int g \, d\mu \geq 0$ implies $g = 0$, μ-a.e., and therefore $\mathbf{z}_i = \mathbf{A}\mathbf{x}_{i-1} + \mathbf{b}_i$, μ-a.e. This is why we do not need to consider the higher-order constraints $\mathbf{M}_r(g\,\mathbf{y}) \succeq 0$ for $r > 0$; only $\mathbf{M}_0(g\,\mathbf{y}) \succeq 0$ ($\Leftrightarrow L_{\mathbf{y}}(g) \geq 0$) suffices. In fact, we impose the stronger linear constraints $L_{\mathbf{y}}(g) = 0$ and $L_{\mathbf{y}}(\mathbf{z}_i - \mathbf{A}\mathbf{x}_{i-1} - \mathbf{b}_i) = 0$ for all $i \in [p]$.

For simplicity, assume that the neural networks have only one single hidden layer, i.e., $m = 1$. Denote by $\mathbf{A}, \mathbf{b}$ the weight and bias respectively. As in (5.3), we use the fact that $l \leq x \leq u$ is equivalent to $(x - l)(x - u) \leq 0$. Then the local Lipschitz constant estimation problem with respect to L_∞-norm can be written as

$$\begin{cases} \sup_{\mathbf{x}, \mathbf{u}, \mathbf{z}, t} & \mathbf{t}^\mathsf{T} \mathbf{A}^\mathsf{T} \text{Diag}(\mathbf{u})\mathbf{c} \\ \text{s.t.} & (\mathbf{z} - \mathbf{A}\mathbf{x} - \mathbf{b})^2 = 0, \quad \mathbf{t}^2 \leq 1 \\ & (\mathbf{x} - \bar{\mathbf{x}}_0 + \varepsilon)(\mathbf{x} - \bar{\mathbf{x}}_0 - \varepsilon) \leq 0 \\ & \mathbf{u}(\mathbf{u} - 1) = 0, \quad (\mathbf{u} - 1/2)\mathbf{z} \geq 0. \end{cases} \qquad \text{(LCEP}_1\text{)}$$

Define the subsets of (LCEP$_1$) to be $I^i = \{x^i, t^i\}$, $J^j = \{u^j, z^j\}$ for $i \in [p_0]$, $j \in [p_1]$, where p_0, p_1 are the number of nodes in the input layer and the hidden layer, respectively. Then the second-order ($r = 2$) heuristic relaxation of (LCEP$_1$) is the following SDP:

$$
\left\{
\begin{aligned}
&\inf_{\mathbf{y}} && L_{\mathbf{y}}(\mathbf{t}^\mathsf{T} \mathbf{A}^\mathsf{T} \operatorname{Diag}(\mathbf{u})\mathbf{c}) \\
&\text{s.t.} && y_0 = 1, \quad \mathbf{M}_1(\mathbf{y}) \succeq 0 \\
& && L_{\mathbf{y}}(\mathbf{z} - \mathbf{A}\mathbf{x} - \mathbf{b}) = 0, \quad L_{\mathbf{y}}((\mathbf{z} - \mathbf{A}\mathbf{x} - \mathbf{b})^2) = 0 \\
& && \mathbf{M}_1(-(x^{(i)} - \bar{x}_0^{(i)} + \varepsilon)(x^{(i)} - \bar{x}_0^{(i)} - \varepsilon)\mathbf{y}, I^i) \succeq 0, \quad i \in [p_0] \quad \text{(HR-2)} \\
& && \mathbf{M}_1((1 - t_i^2)\mathbf{y}, I^i) \succeq 0, \quad \mathbf{M}_2(\mathbf{y}, I^i) \succeq 0, \quad i \in [p_0] \\
& && \mathbf{M}_2(\mathbf{y}, J^j) \succeq 0, \quad \mathbf{M}_1(u_j(u_j - 1)\mathbf{y}, J^j) = 0, \quad j \in [p_1] \\
& && \mathbf{M}_1((u_j - 1/2)z_j\mathbf{y}, J^j) \succeq 0, \quad j \in [p_1].
\end{aligned}
\right.
$$

The r-th order heuristic relaxation (HR-r) also applies to multiple layer neural networks. However, if the neural network has m hidden layers, then the criterion in (5.4) is of degree $m + 1$. If $m \geq 2$, then the first-order moment matrix $\mathbf{M}_1(\mathbf{y})$ is no longer sufficient, as moments of degree > 2 are *not* encoded in $\mathbf{M}_1(\mathbf{y})$ and some may not be encoded in the moment matrices $\mathbf{M}_2(\mathbf{y}, I^i)$, if they include variables of different subsets. See [CLMPa, Appendix E] for more information to deal with higher-degree polynomial objective.

5.4   Overview of numerical experiments

In this section, we provide results for the *global* and *local* Lipschitz constants of *random* networks of fixed size $(80, 80)$ and with various sparsities. We also compute bounds of a *real* trained 1-hidden layer network. For all experiments we focus on the L_∞-norm, the most interesting case for robustness certification. Moreover, we use the Lipschitz constants computed by various methods to certify robustness of a trained network, and compare the ratio of certified inputs with different methods. Let us first provide an overview of the methods with which we compare our results:

- **SHOR**: First-order dense moment relaxation (also called Shor's relaxation) applied to (5.4).

- **HR-1/2**: first/second-order heuristic relaxation applied to (5.4).

- **LP-3/4**: LP-based method, called **LipOpt**, by [LRC20] with degree 3/4, which corresponds to (3.9) without SOS multipliers.

- **LBS**: lower bound obtained by sampling 50000 random points and evaluating the dual norm of the gradient.

The reason why we list **LBS** here is because **LBS** is a valid lower bound on the Lipschitz constant. Therefore all methods should provide a result not lower than **LBS**, a basic necessary condition of consistency.

As discussed earlier, if we want to estimate the global Lipschitz constant, we need the input space Ω to be the whole space. In consideration of numerical issues, we set Ω to be the ball of radius 10 around the origin. For the local Lipschitz constant, we set by default the radius of the input ball as $\varepsilon = 0.1$. In both cases, we compute the Lipschitz constant with respect to the first label. We use the (Python) code provided by [LRC20][1] to execute the experiments for **LipOpt** with the Gurobi solver. For **HR-2** and **SHOR**, we use the YALMIP toolbox (MATLAB) [Lö04] with Mosek as a backend to calculate the Lipschitz constants for *random* networks. For *trained* network, we implement our algorithm in Julia with the Mosek optimizer to accelerate the computation. In the following, running time is referred to the time taken by the LP/SDP solver (Gurobi/Mosek) and "-" means running out of memory during solving the LP/SDP model.

In [LRC20] a certain sparsity structure arising from a neural network was exploited. Consider a neural network F with one single hidden layer, and 4 nodes in each layer. The network F is said to have a sparsity $\omega = 4$ if its weight matrix **A** is symmetric with diagonal blocks of size at most 2×2:

$$\begin{bmatrix} * & * & 0 & 0 \\ * & * & * & 0 \\ 0 & * & * & * \\ 0 & 0 & * & * \end{bmatrix}. \tag{5.6}$$

Larger sparsity values refer to symmetric matrices with band structure of a given size. This sparsity structure (5.6) of the networks greatly influences the number of variables involved in the LP program to solve in [LRC20]. This is in deep contrast with our method which does not require the weight matrix to be as in (5.6). Hence when the network is fully-connected, our method is more efficient and provides tighter upper bounds.

5.4.1 Lipschitz constant estimation

Random networks.   Table 5.1 gives a brief comparison outlook of the results obtained by our method and the method in [LRC20]. For $(80, 80)$ networks, apart from $\omega = 20$, which is not significative, **HR-2** obtains much better bounds and is also much more efficient than **LP-3**. **LP-4** provides tighter bounds than **HR-2** but suffers more computational time, and run out of memory when the sparsity increases. For $(40, 40, 10)$ networks, **HR-1** is a trade-off between **LP-3** and **LP-4**, it provides tighter (resp. looser)

---

[1]https://openreview.net/forum?id=rJe4_xSFDB.

Table 5.1: Bounds of global Lipschitz constants (opt) and solver running time (in seconds) of networks of size $(80, 80)$ and $(40, 40, 10)$.

		(80,80)					(40,40,10)			
		$\omega = 20$	$\omega = 40$	$\omega = 60$	$\omega = 80$		$\omega = 20$	$\omega = 40$	$\omega = 60$	$\omega = 80$
HR-2	opt	1.45	2.05	2.41	2.68	HR-1	0.50	1.16	1.82	2.05
	time	3.14	7.78	8.61	9.82		271.34	165.68	174.86	174.02
LP-3	opt	1.55	2.86	3.85	4.68	LP-3	0.56	1.68	3.01	3.57
	time	2.44	10.36	20.99	71.49		3.84	4.83	7.91	6.33
LP-4	opt	1.43	-	-	-	LP-4	0.29	0.85	-	-
	time	127.99	-	-	-		321.89	28034.27	-	-
LBS	opt	1.05	1.56	1.65	1.86	LBS	0.20	0.48	0 61	0.62

Table 5.2: Comparison of bounds of global Lipschitz constants and solver running time on trained network SDP-NN obtained by **HR-2**, **SHOR**, **LP-3** and **LBS**. The network is a fully connected neural network with one hidden layer, with 784 nodes in the input layer and 500 nodes in the hidden layer. The network is for 10-classification, and we calculate the upper bound with respect to label 2.

	Global				Local			
	HR-2	**SHOR**	**LP-3**	**LBS**	**HR-2**	**SHOR**	**LP-3**	**LBS**
opt	14.56	17.85	-	9.69	12.70	16.07	-	8.20
time	12246	2869	-		20596	4217	-	

bounds than **LP-3** (resp. **LP-4**), but takes more (resp. less) computational time.

Trained Network. Here we use the MNIST classifier SDP-NN described in [RSL18a]. The network is of size $(784, 500)$. In Table 5.2, we see that the **LP-3** algorithm runs out of memory when applied to the real network SDP-NN to compute the global Lipschitz bound. In contrast, **SHOR** and **HR-2** still work and moreover, **HR-2** provides tighter upper bounds than **SHOR** in both global and local cases. As a trade-off, the running time of **HR-2** is around 5 times longer than that of **SHOR**.

5.4.2 Robustness certification

Multi-Classifier. The above SDP-NN network is a well-trained $(784, 500)$ network to classify the digit images from 0 to 9. Denote the parameters of this network by $\mathbf{A} \in \mathbb{R}^{500 \times 784}, \mathbf{b}_1 \in \mathbb{R}^{500}, \mathbf{C} \in \mathbb{R}^{10 \times 500}, \mathbf{b}_2 \in \mathbb{R}^{10}$. The score of an input $\mathbf{x}$ is denoted by $\mathbf{y}^{\mathbf{x}}$, i.e., $\mathbf{y}^{\mathbf{x}} = \mathbf{C} \cdot \mathrm{ReLU}(\mathbf{A}\mathbf{x}_0 + \mathbf{b}_1) + \mathbf{b}_2$.

The label of $\mathbf{x}$, denoted by $r^{\mathbf{x}}$, is the index with the largest score, i.e., $r^{\mathbf{x}} = \arg\max \mathbf{y}^{\mathbf{x}}$. Suppose an input $\mathbf{x}_0$ has label r. For ε and $\mathbf{x}$ such that $\|\mathbf{x} - \mathbf{x}_0\|_\infty \leq \varepsilon$, if for all $i \neq r, y_i^{\mathbf{x}} - y_r^{\mathbf{x}} < 0$, then $\mathbf{x}_0$ is ε-robust. Alternatively, denote by $L_{i,r}^{\mathbf{x}_0,\varepsilon}$ the local Lipschitz constant of function $f_{i,r}(\mathbf{x}) = y_i^{\mathbf{x}} - y_r^{\mathbf{x}}$ with respect to L_∞-norm in the ball $\{\mathbf{x} : \|\mathbf{x} - \mathbf{x}_0\|_\infty \leq \varepsilon\}$. Then the point $\mathbf{x}_0$ is ε-robust if for all $i \neq r, f_{i,r}(\mathbf{x}_0) + \varepsilon L_{i,r}^{\mathbf{x}_0,\varepsilon} < 0$. Since the 28×28 MNIST images are flattened and normalized into vectors taking value in $[0,1]$, we compute the local Lipschitz constant (by **HR-2**) with respect to $\mathbf{x}_0 = \mathbf{0}$ and $\varepsilon = 2$, the complete value is referred to the following matrix:

$$
\mathbf{L} =
\begin{bmatrix}
* & 7.94 & 7.89 & 8.28 & 8.64 & 8.10 & 7.66 & 8.04 & 7.46 & 8.14 \\
7.94 & * & 7.74 & 7.36 & 7.68 & 8.81 & 8.06 & 7.55 & 7.36 & 8.66 \\
7.89 & 7.74 & * & 7.63 & 8.81 & 10.23 & 8.18 & 8.13 & 7.74 & 9.08 \\
8.28 & 7.36 & 7.63 & * & 8.52 & 7.74 & 9.47 & 8.01 & 7.37 & 7.96 \\
8.64 & 7.68 & 8.81 & 8.52 & * & 9.44 & 7.98 & 8.65 & 8.49 & 7.47 \\
8.10 & 8.81 & 10.23 & 7.74 & 9.44 & * & 8.26 & 9.26 & 8.17 & 8.55 \\
7.66 & 8.06 & 8.18 & 9.47 & 7.98 & 8.26 & * & 10.18 & 8.00 & 9.83 \\
8.04 & 7.55 & 8.13 & 8.01 & 8.65 & 9.26 & 10.18 & * & 8.28 & 7.65 \\
7.46 & 7.36 & 7.74 & 7.37 & 8.49 & 8.17 & 8.00 & 8.28 & * & 7.87 \\
8.14 & 8.66 & 9.08 & 7.96 & 7.47 & 8.55 & 9.83 & 7.65 & 7.87 & *
\end{bmatrix}
$$

where $\mathbf{L} = (L_{ij})_{i\neq j}$. Note that if we replace the vector $\mathbf{c}$ in (5.4) by $-\mathbf{c}$, the problem is equivalent to the original one. Therefore, the matrix $\mathbf{L}$ is symmetric, and we only need to compute 45 Lipschitz constants (the upper triangle of $\mathbf{L}$).

Figure 5.4 shows several certified and non-certified examples taken from the MNIST test dataset.

Figure 5.4: Examples of certified points (above) and non-certified points (bellow).

We take different values of ε from 0.02 to 0.1, and compute the ratio of certified examples among the 10000 MNIST test data by the Lipschitz constants we obtain, as shown in Table 5.3. Note that for $\varepsilon = 0.1$, we improve a little bit by 67% compared to **Grad-cert** (65%) described in [RSL18a], as we use an exact formulation of the derivative of ReLU function.

Table 5.3: Ratios of certified test examples for SDP-NN network by **HR-2**.

ε	0.02	0.04	0.06	0.08	0.1
Ratios	97.24%	92.84%	87.10%	78.34%	67.63%

5.5 Notes and sources

Various optimization frameworks have been recently used to certify the robustness of deep neural networks, including SDP [RSL18b, FMP20], LP [DSG+18, WK18], mixed integer programming (MIP) [TXT17], outer polytope approximation [BWC+19, ZWC+18, WZC+18b, WZC+18a], and averaged activation operators [CP20].

Upper bounds on Lipschitz constants of deep networks can be obtained by a product of the layer-wise Lipschitz constants [HCC18]. This is however extremely loose and has many limitations. Note that [VS18] proposed an improvement via a finer product. Departing from this approach, [LRC20] proposed a quadratically constrained quadratic program (QCQP) formulation to estimate the Lipschitz constant of neural networks. Shor's relaxation allows to obtain a valid upper bound. Alternatively, using the LP hierarchy, [LRC20] obtains tighter upper bounds. By another SDP-based method, [FRH+19] provides an upper bound of the Lipschitz constant. However this method is restricted to the L_2-norm whereas most robustness certification problems in deep learning are rather concerned with the L_∞-norm. The heuristic approach presented in this chapter can also apply to nearly sparse polynomial optimization problems, coming from the general optimization literature [CLMP22], or for other types of networks, such as monotone deep equilibrium networks [CLMPb].

The proof of Theorem 5.1 is available in [CLMPa, Appendix 1]. The interested reader can find more complete results for global/local Lipschitz constants of both 1-hidden layer and 2-hidden layer networks with various sizes and sparsities in [CLMPa, Appendix F and G].

Bibliography

[BP21]   Jérôme Bolte and Edouard Pauwels. Conservative set valued
            fields, automatic differentiation, stochastic gradient methods
            and deep learning. *Mathematical Programming*, 188(1):19–51,
            2021.

[BWC⁺19]   Akhilan Boopathy, Tsui-Wei Weng, Pin-Yu Chen, Sijia Liu,
            and Luca Daniel. Cnn-cert: An efficient framework for cer-
            tifying robustness of convolutional neural networks. In *Pro-
            ceedings of the AAAI Conference on Artificial Intelligence*, vol-
            ume 33, pages 3240–3247, 2019.

[CLMPa]   T. Chen, J.-B. Lasserre, V. Magron, and E. Pauwels. Semi-
            algebraic Optimization for Bounding Lipschitz Constants of
            ReLU Networks. *Proceeding of Advances in Neural Information
            Processing Systems 33 (NeurIPS 2020)*.

[CLMPb]   T. Chen, J.-B. Lasserre, V. Magron, and E. Pauwels. Semialge-
            braic Representation of Monotone Deep Equilibrium Models
            and Applications to Certification. *Proceeding of Advances in
            Neural Information Processing Systems 34 (NeurIPS 2021)*.

[CLMP22]   Tong Chen, Jean-Bernard Lasserre, Victor Magron, and
            Edouard Pauwels. A sublevel moment-sos hierarchy for
            polynomial optimization. *Computational Optimization and Ap-
            plications*, 81(1):31–66, 2022.

[CP20]   Patrick L Combettes and Jean-Christophe Pesquet. Lipschitz
            certificates for layered network structures driven by aver-
            aged activation operators. *SIAM Journal on Mathematics of
            Data Science*, 2(2):529–557, 2020.

[DSG⁺18]   Krishnamurthy Dvijotham, Robert Stanforth, Sven Gowal,
            Timothy A Mann, and Pushmeet Kohli. A dual approach
            to scalable verification of deep networks. In *UAI*, pages 550–
            559, 2018.

[FMP20]   Mahyar Fazlyab, Manfred Morari, and George J Pappas. Safety verification and robustness analysis of neural networks via quadratic constraints and semidefinite programming. *IEEE Transactions on Automatic Control*, 2020.

[FRH⁺19]   Mahyar Fazlyab, Alexander Robey, Hamed Hassani, Manfred Morari, and George J. Pappas. Efficient and accurate estimation of lipschitz constants for deep neural networks, 2019.

[HCC18]   Todd Huster, Cho-Yu Jason Chiang, and Ritu Chadha. Limitations of the lipschitz constant as a defense against adversarial examples. In *Joint European Conference on Machine Learning and Knowledge Discovery in Databases*, pages 16–29. Springer, 2018.

[KL18]   Sham M Kakade and Jason D Lee. Provably correct automatic sub-differentiation for qualified programs. In *Advances in neural information processing systems*, pages 7125–7135, 2018.

[LRC20]   Fabian Latorre, Paul Rolland, and Volkan Cevher. Lipschitz constant estimation of neural networks via sparse polynomial optimization. In *International Conference on Learning Representations*, 2020.

[Lö04]   J. Löfberg. Yalmip : A toolbox for modeling and optimization in MATLAB. In *Proceedings of the CACSD Conference*, Taipei, Taiwan, 2004.

[RSL18a]   Aditi Raghunathan, Jacob Steinhardt, and Percy Liang. Certified defenses against adversarial examples. In *International Conference on Learning Representations*, 2018.

[RSL18b]   Aditi Raghunathan, Jacob Steinhardt, and Percy Liang. Semidefinite relaxations for certifying robustness to adversarial examples. In *Advances in Neural Information Processing Systems*, pages 10877–10887, 2018.

[TXT17]   Vincent Tjeng, Kai Xiao, and Russ Tedrake. Evaluating robustness of neural networks with mixed integer programming. *arXiv preprint arXiv:1711.07356*, 2017.

[VS18]   Aladin Virmaux and Kevin Scaman. Lipschitz regularity of deep neural networks: Analysis and efficient estimation. In *Advances in Neural Information Processing Systems*, pages 3835–3844, 2018.

[WK18]   Eric Wong and Zico Kolter. Provable defenses against adver-
            sarial examples via the convex outer adversarial polytope. In
            International Conference on Machine Learning, pages 5286–5295.
            PMLR, 2018.

[WZC+18a]   Lily Weng, Huan Zhang, Hongge Chen, Zhao Song, Cho-Jui
            Hsieh, Luca Daniel, Duane Boning, and Inderjit Dhillon. To-
            wards fast computation of certified robustness for relu net-
            works. In *International Conference on Machine Learning*, pages
            5276–5285. PMLR, 2018.

[WZC+18b]   Tsui-Wei Weng, Huan Zhang, Pin-Yu Chen, Jinfeng Yi, Dong
            Su, Yupeng Gao, Cho-Jui Hsieh, and Luca Daniel. Evaluating
            the robustness of neural networks: An extreme value theory
            approach. *arXiv preprint arXiv:1801.10578*, 2018.

[ZWC+18]   Huan Zhang, Tsui-Wei Weng, Pin-Yu Chen, Cho-Jui Hsieh,
            and Luca Daniel. Efficient neural network robustness certifi-
            cation with general activation functions. In *Advances in neural
            information processing systems*, pages 4939–4948, 2018.

Chapter 6

Noncommutative optimization and quantum information

In this chapter, we handle a specific class of sparse POPs with noncommuting variables, and adapt the concept of CS from Chapter 3 in this setting. A converging hierarchy of semidefinite relaxations for eigenvalue optimization is provided. The Gelfand-Naimark-Segal (GNS) construction is applied to extract optimizers if flatness and irreducibility conditions are satisfied. Among the main techniques used are amalgamation results from operator algebra. The theoretical results are utilized to compute lower bounds on minimal eigenvalue of noncommutative polynomials from the literature, in particular arising from quantum information theory.

6.1 Noncommutative polynomials

We consider a finite alphabet $x_1, \ldots, x_n$ (called noncommuting variables) and generate all possible words (monomials) of finite length in these letters. The empty word is denoted by 1. The resulting set of words is $\langle \underline{x} \rangle$, with $\underline{x} = (x_1, \ldots, x_n)$. We denote by $\mathbb{R}\langle \underline{x} \rangle$ the ring of real polynomials in the noncommutating variables $\underline{x}$. An element in $\mathbb{R}\langle \underline{x} \rangle$ is called a *noncommutative (nc) polynomial*. The *support* of an nc polynomial $f = \sum_{w \in \langle \underline{x} \rangle} a_w w$ is defined by $\mathrm{supp}(f) := \{w \in \langle \underline{x} \rangle \mid a_w \neq 0\}$ and the *degree* of f, denoted by $\deg(f)$, is the length of the longest word in $\mathrm{supp}(f)$. The set of nc polynomials of degree at most r is denoted by $\mathbb{R}\langle \underline{x} \rangle_r$. Let us denote by $\mathbf{W}_r$ the vector of all words of degree at most r with respect to the lexicographic order. Note that $\mathbf{W}_r$ serves as a monomial basis of $\mathbb{R}\langle \underline{x} \rangle_r$ and the length of $\mathbf{W}_r$ is equal to $\sigma(n,r) := \sum_{i=0}^{r} n^i = \frac{n^{r+1}-1}{n-1}$. The ring $\mathbb{R}\langle \underline{x} \rangle$ is equipped with the involution $\star$ that fixes $\mathbb{R} \cup \{x_1, \ldots, x_n\}$ point-wise and reverses words, so that $\mathbb{R}\langle \underline{x} \rangle$ is the $\star$-algebra freely generated by n symmetric letters $x_1, \ldots, x_n$. For instance $(x_1 x_2 + x_2^2 + 1)^{\star} = x_2 x_1 + x_2^2 + 1$. The set of all *symmetric elements* is defined as $\operatorname{Sym} \mathbb{R}\langle \underline{x} \rangle := \{f \in \mathbb{R}\langle \underline{x} \rangle \mid f = f^{\star}\}$. A simple example of element of $\operatorname{Sym} \mathbb{R}\langle \underline{x} \rangle$ is $x_1 x_2 + x_2 x_1 + x_2^2 + 1$. An nc polynomial of the form $g^{\star} g$ is called an *hermitian square*. A given $f \in \operatorname{Sym} \mathbb{R}\langle \underline{x} \rangle$ is a sum of Hermitian squares (SOHS) if there exist nc polynomials $h_1, \ldots, h_t \in \mathbb{R}\langle \underline{x} \rangle$ such that $f = h_1^{\star} h_1 + \cdots + h_t^{\star} h_t$. Let $\Sigma\langle \underline{x} \rangle$ stand for the set of SOHS. We denote by $\Sigma\langle \underline{x} \rangle_r \subseteq \Sigma\langle \underline{x} \rangle$ the set of SOHS polynomials of degree at most $2r$. We now recall how to check whether a given $f \in \operatorname{Sym} \mathbb{R}\langle \underline{x} \rangle$ is an SOHS. The existing procedure, known as the *Gram matrix method*, relies on the following proposition.

Proposition 6.1 *Assume that $f \in \operatorname{Sym} \mathbb{R}\langle \underline{x} \rangle$ is of degree at most $2d$. Then $f \in \Sigma\langle \underline{x} \rangle$ if and only if there exists $\mathbf{G}_f \succeq 0$ satisfying*

$$f = \mathbf{W}_d^{\star} \mathbf{G}_f \mathbf{W}_d. \tag{6.1}$$

Conversely, given such $\mathbf{G}_f \succeq 0$ of rank t, one can construct $g_1, \ldots, g_t \in \mathbb{R}\langle \underline{x} \rangle$ of degree at most d such that $f = \sum_{i=1}^{t} g_i^{\star} g_i$.

Any symmetric matrix $\mathbf{G}_f$ (not necessarily positive semidefinite) satisfying (6.1) is called a *Gram matrix* of f.

Given a set of nc polynomials $\mathfrak{g} = \{g_1, \ldots, g_m\} \subseteq \operatorname{Sym} \mathbb{R}\langle \underline{x} \rangle$, the nc semialgebraic set $\mathcal{D}_{\mathfrak{g}}$ associated to $\mathfrak{g}$ is defined as follows:

$$\mathcal{D}_{\mathfrak{g}} := \bigcup_{k \in \mathbb{N}^*} \{\underline{A} = (A_1, \ldots, A_n) \in (\mathbb{S}_k)^n \mid g_j(\underline{A}) \succeq 0, j \in [m]\}. \tag{6.2}$$

When considering only tuples of $k \times k$ symmetric matrices, we use the notation $\mathcal{D}_{\mathfrak{g}}^k := \mathcal{D}_{\mathfrak{g}} \cap (\mathbb{S}_k)^n$. The operator semialgebraic set $\mathcal{D}_{\mathfrak{g}}^{\infty}$ is the set of all bounded self-adjoint operators $\underline{A}$ on a Hilbert space $\mathcal{H}$ endowed with a scalar product $\langle \cdot \mid \cdot \rangle$, making $g(\underline{A})$ a positive semidefinite operator for all $g \in \mathfrak{g}$, i.e., $\langle g(\underline{A})v \mid v \rangle \geq 0$, for all $v \in \mathcal{H}$. We say that an nc polynomial f is positive (denoted by $f \succ 0$) on $\mathcal{D}_{\mathfrak{g}}^{\infty}$ if for all $\underline{A} \in \mathcal{D}_{\mathfrak{g}}^{\infty}$ the operator $f(\underline{A})$ is positive definite, i.e., $\langle f(\underline{A})v \mid v \rangle > 0$, for all nonzero $v \in \mathcal{H}$. The quadratic module $\mathcal{M}(\mathfrak{g})$, generated by $\mathfrak{g}$, is defined by

$$\mathcal{M}(\mathfrak{g}) := \left\{ \sum_{i=1}^{t} a_i^{\star} g_i a_i \mid t \in \mathbb{N}^*, a_i \in \mathbb{R}\langle \underline{x} \rangle, g_i \in \mathfrak{g} \cup \{1\} \right\}. \tag{6.3}$$

Given $r \in \mathbb{N}^*$, the truncated quadratic module $\mathcal{M}(\mathfrak{g})_r$ of order r, generated by $\mathfrak{g}$, is

$$\mathcal{M}(\mathfrak{g})_r := \left\{ \sum_{i=1}^{t} a_i^{\star} g_i a_i \mid t \in \mathbb{N}^*, a_i \in \mathbb{R}\langle \underline{x} \rangle, g_j \in \mathfrak{g} \cup \{1\}, \deg(a_i^{\star} g_i a_i) \leq 2r \right\}.$$

$$\tag{6.4}$$

A quadratic module $\mathcal{M}$ is said to be *Archimedean* if for each $a \in \mathbb{R}\langle \underline{x} \rangle$, there exists $N > 0$ such that $N - a^{\star}a \in \mathcal{M}$. One can show that this is equivalent to the existence of an $N > 0$ such that $N - \sum_{i=1}^{n} x_i^2 \in \mathcal{M}$.

We also recall the nc analog of Putinar's Positivstellensatz (Theorem 2.3) describing nc polynomials positive on $\mathcal{D}_\mathfrak{g}^\infty$ with Archimedean $\mathcal{M}(\mathfrak{g})$.

Theorem 6.2 (Helton-McCullough) *Let* $\{f\} \cup \mathfrak{g} \subseteq \mathrm{Sym}\,\mathbb{R}\langle \underline{x} \rangle$ *and assume that* $\mathcal{M}(\mathfrak{g})$ *is Archimedean. If* $f(\underline{A}) \succ 0$ *for all* $\underline{A} \in \mathcal{D}_\mathfrak{g}^\infty$, *then* $f \in \mathcal{M}(\mathfrak{g})$.

6.2 Correlative sparsity patterns

We rely on the same CS framework as in Chapter 3. More concretely, assuming $f = \sum_w a_w w \in \mathrm{Sym}\,\mathbb{R}\langle \underline{x} \rangle$ and $\mathfrak{g} = \{g_1, \ldots, g_m\} \subseteq \mathrm{Sym}\,\mathbb{R}\langle \underline{x} \rangle$, we define the csp graph associated with f and $\mathfrak{g}$ to be the graph G^{csp} with nodes $V = [n]$ and with edges E satisfying $\{i, j\} \in E$ if one of following conditions holds:

(i) there exists $w \in \mathrm{supp}(f)$ s.t. $x_i, x_j \in \mathrm{var}(w)$;

(ii) there exists $k \in [m]$ s.t. $x_i, x_j \in \mathrm{var}(g_k)$,

where we use $\mathrm{var}(g)$ to denote the set of variables effectively involved in $g \in \mathbb{R}\langle \underline{x} \rangle$. Let $(G^{\mathrm{csp}})'$ be a chordal extension of G^{csp} and $I_k, k \in [p]$ be the maximal cliques of $(G^{\mathrm{csp}})'$ with cardinality being denoted by $n_k, k \in [p]$. We denote by $\langle \underline{x}(I_k) \rangle$ (resp. $\mathbb{R}\langle \underline{x}, I_k \rangle$) the set of words (resp. nc polynomials) in the n_k variables $\underline{x}(I_k) = \{x_i : i \in I_k\}$. We also define $\mathrm{Sym}\,\mathbb{R}\langle \underline{x}, I_k \rangle :=$ $\mathrm{Sym}\,\mathbb{R}\langle \underline{x} \rangle \cap \mathbb{R}\langle \underline{x}, I_k \rangle$. Let $\Sigma\langle \underline{x}, I_k \rangle$ stand for the set of SOHS in $\mathbb{R}\langle \underline{x}, I_k \rangle$ and we denote by $\Sigma\langle \underline{x}, I_k \rangle_r$ the restriction of $\Sigma\langle \underline{x}, I_k \rangle$ to nc polynomials of degree at most $2r$. In the sequel, we will rely on two specific assumptions. The first one is as follows.

Assumption 6.3 (Boundedness) *Let* $\mathcal{D}_\mathfrak{g}$ *be as in* (6.2). *There exists* $N > 0$ *such that* $\sum_{i=1}^{n} x_i^2 \preceq N$, *for all* $\underline{x} \in \mathcal{D}_\mathfrak{g}^\infty$.

Then, Assumption 6.3 implies that $\sum_{j \in I_k} x_j^2 \preceq N$, for all $k \in [p]$. Thus we define

$$g_{m+k} := N - \sum_{j \in I_k} x_j^2, \quad k \in [p], \tag{6.5}$$

and set $m' = m + p$ in order to describe the same set $\mathcal{D}_\mathfrak{g}$ again as

$$\mathcal{D}_\mathfrak{g} := \bigcup_{k \in \mathbb{N}^*} \{\underline{A} \in (\mathbb{S}_k)^n \mid g_j(\underline{A}) \succeq 0, j \in [m']\}, \tag{6.6}$$

as well as the operator semialgebraic set $\mathcal{D}_\mathfrak{g}^\infty$.

The second assumption, which is the strict nc analog of Assumption 3.1(i)–(iii), is as follows.

Assumption 6.4 *Let $\mathcal{D}_g$ be as in (6.6) and let $f \in \mathrm{Sym}\,\mathbb{R}\langle\underline{x}\rangle$. The index set $J := \{1,\dots,m'\}$ is partitioned into p disjoint sets $J_1,\dots,J_p$ and the two collections $\{I_1,\dots,I_p\}$ and $\{J_1,\dots,J_p\}$ satisfy*

(i) *The objective function f can be decomposed as $f = f_1 + \cdots + f_p$, with $f_k \in \mathrm{Sym}\,\mathbb{R}\langle\underline{x},I_k\rangle$ for all $k \in [p]$;*

(ii) *For all $k \in [p]$ and $j \in J_k$, $g_j \in \mathrm{Sym}\,\mathbb{R}\langle\underline{x},I_k\rangle$;*

(iii) *The RIP (1.8) holds for $I_1,\dots,I_p$ (possibly after some reordering).*

6.3   Noncommutative moment and localizing matrices

Given a sequence $\mathbf{y} = (y_w)_{w\in\mathbf{W}_{2r}} \in \mathbb{R}^{\sigma(n,2r)}$ (here we allow $r = \infty$), let us define the linear functional $L_\mathbf{y} : \mathbb{R}\langle\underline{x}\rangle_{2r} \to \mathbb{R}$ by $L_\mathbf{y}(f) := \sum_w a_w y_w$, for every polynomial $f = \sum_w a_w w$ of degree at most $2r$. The sequence $\mathbf{y}$ is said to be *unital* if $y_1 = 1$ and is said to be *symmetric* if $y_{w^\star} = y_w$ for all $w \in \mathbf{W}_{2r}$. Suppose $g \in \mathrm{Sym}\,\mathbb{R}\langle\underline{x}\rangle$ with $\deg(g) \leq 2r$. We further associate to $\mathbf{y}$ the following two matrices:

(1) the *(noncommutative) moment matrix* $\mathbf{M}_r(\mathbf{y})$ is the matrix indexed by words $u,v \in \mathbf{W}_r$, with $[\mathbf{M}_r(\mathbf{y})]_{u,v} = L_\mathbf{y}(u^\star v) = y_{u^\star v}$;

(2) the *localizing matrix* $\mathbf{M}_{r-\lceil\deg(g)/2\rceil}(g\mathbf{y})$ is the matrix indexed by words $u,v \in \mathbf{W}_{r-\lceil\deg(g)/2\rceil}$, with $[\mathbf{M}_{r-\lceil\deg(g)/2\rceil}(g\mathbf{y})]_{u,v} = L_\mathbf{y}(u^\star g v)$.

We recall the following useful facts.

Lemma 6.5 *Let $g \in \mathrm{Sym}\,\mathbb{R}\langle\underline{x}\rangle$ with $\deg(g) \leq 2r$ and let L be the linear functional associated to a symmetric sequence $\mathbf{y} := (y_w)_{w\in\mathbf{W}_{2r}} \in \mathbb{R}^{\sigma(n,2r)}$. Then,*

(1) $L_\mathbf{y}(h^\star h) \geq 0$ *for all $h \in \mathbb{R}\langle\underline{x}\rangle_r$ if and only if the moment matrix $\mathbf{M}_r(\mathbf{y}) \succeq 0$;*

(2) $L_\mathbf{y}(h^\star g h) \geq 0$ *for all $h \in \mathbb{R}\langle\underline{x}\rangle_{r-\lceil\deg(g)/2\rceil}$ if and only if the localizing matrix $\mathbf{M}_{r-\lceil\deg(g)/2\rceil}(g\mathbf{y}) \succeq 0$.*

Definition 6.6 *Let $\mathbf{y} = (y_w)_{w\in\mathbf{W}_{2r+2\delta}} \in \mathbb{R}^{\sigma(n,2r+2\delta)}$ and $\tilde{\mathbf{y}} = (y_w)_{w\in\mathbf{W}_{2r}}$ be its truncation. We can write the moment matrix $\mathbf{M}_{r+\delta}(\mathbf{y})$ in block form:*

$$\mathbf{M}_{r+\delta}(\mathbf{y}) = \begin{bmatrix} \mathbf{M}_r(\tilde{\mathbf{y}}) & B \\ B^\mathsf{T} & C \end{bmatrix}.$$

We say that $\mathbf{y}$ is δ-flat or that $\mathbf{y}$ is a flat extension of $\tilde{\mathbf{y}}$, if $\mathbf{M}_{r+\delta}(\mathbf{y})$ is flat over $\mathbf{M}_r(\tilde{L})$, i.e., if $\mathrm{rank}\,\mathbf{M}_{r+\delta}(\mathbf{y}) = \mathrm{rank}\,\mathbf{M}_r(\tilde{\mathbf{y}})$.

For a subset $I \subseteq [n]$, let us define $\mathbf{M}_r(\mathbf{y}, I)$ to be the moment submatrix obtained from $\mathbf{M}_r(\mathbf{y})$ after retaining only those rows and columns indexed by $w \in \langle \underline{x}(I) \rangle_r$. For $g \in \mathbb{R}\langle \underline{x}, I \rangle$ with $\deg(g) \leq 2r$, we also define the localizing submatrix $\mathbf{M}_{r - \lceil \deg(g)/2 \rceil}(g\mathbf{y}, I)$ in a similar fashion.

6.4   Sparse representations

Here, we state our main theoretical result, which is a sparse version of the Helton-McCullough Positivstellensatz (Theorem 6.2). For this, we rely on amalgamation theory for C^*-algebras.

Given a Hilbert space $\mathcal{H}$, we denote by $\mathcal{B}(\mathcal{H})$ the set of bounded operators on $\mathcal{H}$. A C^*-algebra is a complex Banach algebra $\mathcal{A}$ (thus also a Banach space), endowed with a norm $\| \cdot \|$, and with an involution $\star$ satisfying $\| xx^* \| = \| x \|^2$ for all $x \in \mathcal{A}$. Equivalently, it is a norm closed subalgebra with involution of $\mathcal{B}(\mathcal{H})$ for some Hilbert space $\mathcal{H}$. Given a C^*-algebra $\mathcal{A}$, a *state* φ is defined to be a positive linear functional of unit norm on $\mathcal{A}$, and we write often $(\mathcal{A}, \varphi)$ when $\mathcal{A}$ comes together with the state φ. Given two C^*-algebras $(\mathcal{A}_1, \varphi_1)$ and $(\mathcal{A}_2, \varphi_2)$, a homomorphism $\iota : \mathcal{A}_1 \to \mathcal{A}_2$ is called *state-preserving* if $\varphi_2 \circ \iota = \varphi_1$. Given a C^*-algebra $\mathcal{A}$, a *unitary representation* of $\mathcal{A}$ in $\mathcal{H}$ is a $\star$-homomorphism $\pi : \mathcal{A} \to \mathcal{B}(\mathcal{H})$ which is *strongly continuous*, i.e., the mapping $\mathcal{A} \to \mathcal{H}$, $g \mapsto \pi(g)\xi$ is continuous for every $\xi \in \mathcal{H}$.

Theorem 6.7 *Let $(\mathcal{A}, \varphi_0)$ and $\{ (\mathcal{B}_k, \varphi_k) : k \in I \}$ be C^*-algebras with states, and let ι_k be a state-preserving embedding of $\mathcal{A}$ into $\mathcal{B}_k$, for each $k \in I$. Then there exists a C^*-algebra $\mathcal{C}$ amalgamating the $(\mathcal{B}_k, \varphi_k)$ over $(\mathcal{A}, \varphi_0)$. That is, there is a state φ on $\mathcal{C}$, and state-preserving homomorphisms $j_k : \mathcal{B}_k \to \mathcal{C}$, such that $j_k \circ \iota_k = j_i \circ \iota_i$, for all $k, i \in I$, and such that $\bigcup_{k \in I} j_k(\mathcal{B}_k)$ generates $\mathcal{C}$.*

Theorem 6.7 is illustrated in Figure 6.1 in the case $I = \{1, 2\}$. We also recall the Gelfand-Naimark-Segal (GNS) construction establishing a correspondence between $\star$-representations of a C^*-algebra and positive linear functionals on it. In our context, the next result restricts to linear functionals on $\mathbb{R}\langle \underline{x} \rangle$ which are positive on an Archimedean quadratic module.

Theorem 6.8 *Let $\mathfrak{g} \subseteq \mathrm{Sym}\, \mathbb{R}\langle \underline{x} \rangle$ be given such that its quadratic module $\mathcal{M}(\mathfrak{g})$ is Archimedean. Let $L : \mathbb{R}\langle \underline{x} \rangle \to \mathbb{R}$ be a nontrivial linear functional with $L(\mathcal{M}(\mathfrak{g})) \subseteq \mathbb{R}_{\geq 0}$. Then there exists a tuple $\underline{A} = (A_1, \ldots, A_n) \in \mathcal{D}_{\mathfrak{g}}^{\infty}$ and a vector $\mathbf{v}$ such that $L(f) = \langle f(\underline{A})\mathbf{v}, \mathbf{v} \rangle$, for all $f \in \mathbb{R}\langle \underline{x} \rangle$.*

Let $I_k, k \in [p]$ and $J_k, k \in [p]$ be given as in Section 6.2. For $k \in [p]$, let us define

$$\mathcal{M}(\mathfrak{g})^k := \left\{ a_0^* a_0 + \sum_{i \in J_k} a_i^* g_i a_i \;\middle|\; a_i \in \mathbb{R}\langle \underline{x}, I_k \rangle, i \in J_k \cup \{0\} \right\}$$

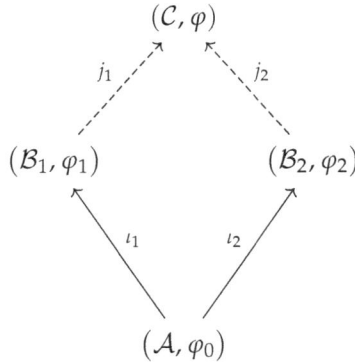

Figure 6.1: Illustration of Theorem 6.7 in the case $I = \{1,2\}$.

and

$$\mathcal{M}(\mathfrak{g})^{\mathrm{cs}} := \mathcal{M}(\mathfrak{g})^1 + \cdots + \mathcal{M}(\mathfrak{g})^p. \tag{6.7}$$

Next, we state the main foundational result of this section.

Theorem 6.9 *Let $\{f\} \cup \mathfrak{g} \subseteq \mathrm{Sym}\, \mathbb{R}\langle \underline{x} \rangle$ and let $\mathcal{D}_{\mathfrak{g}}$ be as in (6.6) with the additional quadratic constraints (6.5). Suppose Assumption 6.4 holds. If $f(\underline{A}) \succ 0$ for all $\underline{A} \in \mathcal{D}_{\mathfrak{g}}^{\infty}$, then $f \in \mathcal{M}(\mathfrak{g})^{\mathrm{cs}}$.*

We provide an example demonstrating that sparsity without an RIP-type condition is not sufficient to deduce sparsity in SOHS decompositions.

Example 6.10 *Consider the case of three variables $\underline{x} = (x_1, x_2, x_3)$ and the polynomial*

$$\begin{aligned}
f &= (x_1 + x_2 + x_3)^2 \\
&= x_1^2 + x_2^2 + x_3^2 + x_1 x_2 + x_2 x_1 + x_1 x_3 + x_3 x_1 + x_2 x_3 + x_3 x_2 \in \Sigma\langle \underline{x} \rangle.
\end{aligned}$$

Then $f = f_1 + f_2 + f_3$, with

$$f_1 = \frac{1}{2}x_1^2 + \frac{1}{2}x_2^2 + x_1 x_2 + x_2 x_1 \in \mathbb{R}\langle x_1, x_2 \rangle,$$

$$f_2 = \frac{1}{2}x_2^2 + \frac{1}{2}x_3^2 + x_2 x_3 + x_3 x_2 \in \mathbb{R}\langle x_2, x_3 \rangle,$$

$$f_3 = \frac{1}{2}x_1^2 + \frac{1}{2}x_3^2 + x_1 x_3 + x_3 x_1 \in \mathbb{R}\langle x_1, x_3 \rangle.$$

However, the sets $I_1 = \{1,2\}$, $I_2 = \{2,3\}$ and $I_3 = \{1,3\}$ do not satisfy the RIP condition (1.8) and $f \notin \Sigma\langle \underline{x} \rangle^{cs} := \Sigma\langle x_1, x_2 \rangle + \Sigma\langle x_2, x_3 \rangle + \Sigma\langle x_1, x_3 \rangle$ since it has a unique Gram matrix by homogeneity.

Now consider $\mathfrak{g} = \{1 - x_1^2, 1 - x_2^2, 1 - x_3^2\}$. Then $\mathcal{D}_{\mathfrak{g}}$ is as in (6.6), $\mathcal{M}(\mathfrak{g})^{cs}$ is as in (6.7) and $f|_{\mathcal{D}_{\mathfrak{g}}^\infty} \succeq 0$. However, we claim that $f - b \in \mathcal{M}(\mathfrak{g})^{cs}$ if and only if $b \leq -3$. Clearly,

$$f + 3 = (x_1 + x_2)^2 + (x_1 + x_3)^2 + (x_2 + x_3)^2$$
$$+ (1 - x_1^2) + (1 - x_2^2) + (1 - x_3^2) \in \mathcal{M}(\mathfrak{g})^{cs}.$$

So one has $-3 \leq \sup\{b : f - b \in \mathcal{M}(\mathfrak{g})^{cs}\}$, and the dual of this latter problem is given by

$$\begin{cases} \inf_{\mathbf{y}_k} \ \sum_{k=1}^3 L_{\mathbf{y}_k}(f_k) \\ \text{s.t.} \ \ L_{\mathbf{y}_k}(1) = 1, \quad k = 1,2,3 \\ \quad\quad L_{\mathbf{y}_k}(h^* h) \succeq 0, \quad \forall h \in \mathbb{R}\langle \underline{x}, I_k \rangle, \quad k = 1,2,3 \\ \quad\quad L_{\mathbf{y}_k}(h^*(1 - x_i^2)h) \succeq 0, \quad \forall h \in \mathbb{R}\langle \underline{x}, I_k \rangle, i \in I_k, k = 1,2,3 \\ \quad\quad L_{\mathbf{y}_j}|_{\mathbb{R}\langle \underline{X}(I_j \cap I_k) \rangle} = L_{\mathbf{y}_k}|_{\mathbb{R}\langle \underline{X}(I_j \cap I_k) \rangle}, \quad j,k = 1,2,3. \end{cases} \quad (6.8)$$

Hence, by weak duality, it suffices to show that there exist linear functionals $L_{\mathbf{y}_k} : \mathbb{R}\langle \underline{x}, I_k \rangle \to \mathbb{R}$ satisfying the constraints of problem (6.8) and such that $\sum_k L_{\mathbf{y}_k}(f_k) = -3$. Define

$$A = \begin{bmatrix} 0 & 1 \\ 1 & 0 \end{bmatrix}, \quad B = -A$$

and let

$$L_{\mathbf{y}_k}(g) = \operatorname{tr}(g(A, B)) \quad \text{for } g \in \mathbb{R}\langle \underline{x}, I_k \rangle.$$

Since $L_{\mathbf{y}_k}(f_k) = -1$, the three first constraints of problem (6.8) are easily verified and $\sum_k L_{\mathbf{y}_k}(f_k) = -3$. For the last one, given, say $h \in \mathbb{R}\langle \underline{x}, I_1 \rangle \cap \mathbb{R}\langle \underline{x}, I_2 \rangle = \mathbb{R}\langle x_2 \rangle$, we have

$$L_{\mathbf{y}_1}(h) = \operatorname{tr}(h(B)),$$
$$L_{\mathbf{y}_2}(h) = \operatorname{tr}(h(A)),$$

since $L_{\mathbf{y}_1}$ (resp. $L_{\mathbf{y}_2}$) is defined on $\mathbb{R}\langle x_1, x_2 \rangle$ (resp. $\mathbb{R}\langle x_2, x_3 \rangle$) and h depends only on the second (resp. first) variable x_2 corresponding to B (resp. A).

But matrices A and B are orthogonally equivalent as $UAU^\mathsf{T} = B$ for

$$U = \begin{bmatrix} 0 & 1 \\ -1 & 0 \end{bmatrix},$$

whence $h(B) = h(UAU^\mathsf{T}) = Uh(A)U^\mathsf{T}$ and $h(A)$ have the same trace.

6.5   Sparse GNS construction

Next, we provide the main theoretical tools to extract solutions of nc optimization problems with CS. To this end, we first present sparse nc versions of theorems by Curto and Fialkow. As recalled in Section 2.4 for the commutative case, Curto and Fialkow provided sufficient conditions for linear functionals on the set of degree $2r$ polynomials to be represented by integration with respect to a nonnegative measure. The main sufficient condition to guarantee such a representation is flatness (see Definition 6.6) of the corresponding moment matrix. We recall this result, which relies on a finite-dimensional GNS construction.

Theorem 6.11 *Let $\mathfrak{g} \subseteq \mathrm{Sym}\,\mathbb{R}\langle \underline{x} \rangle$ and set $\delta := \max\{\lceil \deg(g)/2 \rceil : g \in \mathfrak{g}\}$. For $r \in \mathbb{N}^*$, let $L_\mathbf{y} : \mathbb{R}\langle \underline{x} \rangle_{2r+2\delta} \to \mathbb{R}$ be the linear functional associated to a unital sequence $\mathbf{y} = (y_w)_{w \in \mathbf{W}_{2r+2\delta}} \in \mathbb{R}^{\sigma(n,2r+2\delta)}$ satisfying $L_\mathbf{y}(\mathcal{M}(\mathfrak{g})_{r+\delta}) \subseteq \mathbb{R}_{\geq 0}$. If $\mathbf{y}$ is δ-flat, then there exists $\hat{A} \in \mathcal{D}_\mathfrak{g}^t$ for some $t \leq \sigma(n,r)$ and a unit vector $\mathbf{v}$ such that*

$$L_\mathbf{y}(g) = \langle g(\hat{A})\mathbf{v}, \mathbf{v} \rangle, \qquad\qquad (6.9)$$

for all $g \in \mathrm{Sym}\,\mathbb{R}\langle \underline{x} \rangle_{2r}$.

We now give the sparse version of Theorem 6.11.

Theorem 6.12 *Suppose $r \in \mathbb{N}^*$. Let $\mathfrak{g} \subseteq \mathrm{Sym}\,\mathbb{R}\langle \underline{x} \rangle_{2r}$, and assume $\mathcal{D}_\mathfrak{g}$ is as in (6.6) with the additional quadratic constraints (6.5). Suppose Assumption 6.4(i) holds. Set $\delta := \max\{\lceil \deg(g)/2 \rceil : g \in \mathfrak{g}\}$. Let $L_\mathbf{y} : \mathbb{R}\langle \underline{x} \rangle_{2r+2\delta} \to \mathbb{R}$ be the linear functional associated to a unital sequence $\mathbf{y} = (y_w)_{w \in \mathbf{W}_{2r+2\delta}} \in \mathbb{R}^{\sigma(n,2r+2\delta)}$ satisfying $L_\mathbf{y}(\mathcal{M}(\mathfrak{g})_{r+\delta}) \subseteq \mathbb{R}_{\geq 0}$. Assume that the following holds:*

(H1) $\mathbf{M}_{r+\delta}(\mathbf{y}, I_k)$ and $\mathbf{M}_{r+\delta}(\mathbf{y}, I_k \cap I_j)$ are δ-flat, for all $j,k \in [p]$.

Then, there exist finite-dimensional Hilbert spaces $\mathcal{H}(I_k)$ with dimension t_k, for all $k \in [p]$, Hilbert spaces $\mathcal{H}(I_j \cap I_k) \subseteq \mathcal{H}(I_j), \mathcal{H}(I_k)$ for all pairs (j,k) with $I_j \cap I_k \neq 0$, and operators $\hat{A}^k, \hat{A}^{jk}$, acting on them, respectively. Further, there are unit vectors $\mathbf{v}^j \in \mathcal{H}(I_j)$ and $\mathbf{v}^{jk} \in \mathcal{H}(I_j \cap I_k)$ such that

$$\begin{aligned} L_\mathbf{y}(f) &= \langle f(\hat{A}^j)\mathbf{v}^j, \mathbf{v}^j \rangle \quad \text{for all } f \in \mathbb{R}\langle \underline{x}, I_j \rangle_{2r}, \\ L_\mathbf{y}(g) &= \langle g(\hat{A}^{jk})\mathbf{v}^{jk}, \mathbf{v}^{jk} \rangle \quad \text{for all } g \in \mathbb{R}\langle \underline{X}(I_j \cap I_k) \rangle_{2r}. \end{aligned} \qquad (6.10)$$

Assuming that for all pairs (j,k) with $I_j \cap I_k \neq \emptyset$, one has

(H2) *the matrices* $(\hat{A}_i^{jk})_{i \in I_j \cap I_k}$ *have no common complex invariant sub-spaces,*

then there exist $\underline{A} \in \mathcal{D}_{\mathfrak{g}}^t$*, with* $t := t_1 \cdots t_p$*, and a unit vector* **v** *such that*

$$L_{\mathbf{y}}(f) = \langle f(\underline{A})\mathbf{v}, \mathbf{v} \rangle, \qquad (6.11)$$

for all $f \in \sum_k \mathbb{R}\langle \underline{x}, I_k \rangle_{2r}$*.*

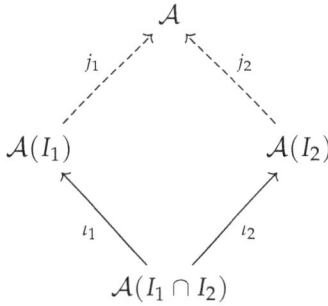

Figure 6.2: Amalgamation of finite-dimensional $C^\star$-algebras.

Example 6.13 (Non-amalgamation in finite-dimensional algebras) *Given* I_1 *and* I_2*, suppose* $\mathcal{A}(I_1 \cap I_2)$ *is generated by the* 2×2 *diagonal matrix*

$$A^{12} = \begin{bmatrix} 1 & \\ & 2 \end{bmatrix},$$

and assume $\mathcal{A}(I_1) = \mathcal{A}(I_2) = \mathbb{M}_3(\mathbb{R})$*. (Observe that* $\mathcal{A}(I_1 \cap I_2)$ *is the algebra of all diagonal matrices.) For each* $k \in \{1, 2\}$*, let us define* $\iota_k(A) := A \oplus k$*, for all* $A \in \mathcal{A}(I_1 \cap I_2)$*. We claim that there is no finite-dimensional* $C^\star$*-algebra* $\mathcal{A}$ *amalgamating the above Figure 6.2. Indeed, by the Skolem-Noether theorem, every homomorphism* $\mathbb{M}_n(\mathbb{R}) \to \mathbb{M}_m(\mathbb{R})$ *is of the form* $x \mapsto P^{-1}(x \otimes \mathbf{I}_{m/n})P$ *for some invertible* P*; in particular, n divides m. If a desired* $\mathcal{A}$ *existed, then the matrices* $(A^{12} \oplus 1) \otimes \mathbf{I}_k$ *and* $(A^{12} \oplus 2) \otimes \mathbf{I}_k$ *would be similar. But they are not as is easily seen from eigenvalue multiplicities.*

As in the dense case, we can summarize the sparse GNS construction procedure described in the proof of Theorem 6.12 into an algorithm, called SparseGNS (see [KMP21, Algorithm 4.6]).

6.6   Eigenvalue optimization

We provide SDP relaxations allowing one to under-approximate the smallest eigenvalue that a given nc polynomial can attain on a tuple of symmetric matrices from a given semialgebraic set. We first recall the celebrated Helton-McCullough theorem stating the equivalence between SOHS and positive semidefinite nc polynomials.

Theorem 6.14 (Helton-McCullough) *Given $f \in \mathrm{Sym}\, \mathbb{R}\langle \underline{x} \rangle$, $f(\underline{A}) \succeq 0$, for all $\underline{A} \in (\mathbb{S}_k)^n$, $k \in \mathbb{N}^*$, if and only if $f \in \Sigma \langle \underline{x} \rangle$.*

In contrast with the constrained case where we obtain the analog of Putinar's Positivstellensatz in Theorem 6.9, there is no sparse analog of Theorem 6.14, as shown in the following example.

Lemma 6.15 *There exist polynomials which are sparse sums of hermitian squares but are not sums of sparse hermitian squares.*

PROOF   Let $v = \begin{bmatrix} x_1 & x_1 x_2 & x_2 & x_3 & x_3 x_2 \end{bmatrix}$,

$$\mathbf{G}_f = \begin{bmatrix} 1 & -1 & -1 & 0 & \alpha \\ -1 & 2 & 0 & -\alpha & 0 \\ -1 & 0 & 3 & -1 & 9 \\ 0 & -\alpha & -1 & 6 & -27 \\ \alpha & 0 & 9 & -27 & 142 \end{bmatrix}, \qquad \alpha \in \mathbb{R}, \qquad (6.12)$$

and consider

$$\begin{aligned} f &= v\mathbf{G}_f v^* \\ &= x_1^2 - x_1 x_2 - x_2 x_1 + 3x_2^2 - 2x_1 x_2 x_1 + 2x_1 x_2^2 x_1 \\ &\quad - x_2 x_3 - x_3 x_2 + 6x_3^2 + 9x_2^2 x_3 + 9x_3 x_2^2 - 54 x_3 x_2 x_3 + 142 x_3 x_2^2 x_3. \end{aligned} \qquad (6.13)$$

The polynomial f is clearly sparse with respect to $I_1 = \{x_1, x_2\}$ and $I_2 = \{x_2, x_3\}$. Note that the matrix $\mathbf{G}_f$ is positive semidefinite if and only if $0.270615 \lesssim \alpha \lesssim 1.1075$, whence f is a sparse polynomial that is an SOHS.
    We claim that $f \notin \Sigma \langle \underline{x}, I_1 \rangle + \Sigma \langle \underline{x}, I_2 \rangle$, i.e., f is not a sum of sparse hermitian squares. By the Newton chip method only monomials in v can appear in an SOHS decomposition of f. Further, every Gram matrix of f in the monomial basis v is of the form (6.12). However, the matrix $\mathbf{G}_f$ with $\alpha = 0$ is not positive semidefinite, and hence $f \notin \Sigma \langle \underline{x}, I_1 \rangle + \Sigma \langle \underline{x}, I_2 \rangle$.

6.6.1   Unconstrained eigenvalue optimization

Let $\mathbf{I}_k$ stand for the $k \times k$ identity matrix. Given $f \in \mathrm{Sym}\, \mathbb{R}\langle \underline{x} \rangle$ of degree $2d$, the smallest eigenvalue of f is obtained by solving the following optimization problem:

$$\lambda_{\min}(f) := \inf \{ \langle f(\underline{A})\mathbf{v}, \mathbf{v} \rangle : \underline{A} \in (\mathbb{S}_k)^n, k \in \mathbb{N}^*, \|\mathbf{v}\|_2 = 1 \}. \qquad (6.14)$$

The optimal value $\lambda_{\min}(f)$ of Problem (6.14) is the greatest lower bound on the eigenvalues of $f(\underline{A})$ over all n-tuples $\underline{A}$ of real symmetric matrices. Problem (6.14) can be rewritten as follows:

$$\lambda_{\min}(f) = \sup_{b} \quad b$$
$$\text{s.t.} \quad f(\underline{A}) - b\mathbf{I}_k \succeq 0, \quad \forall \underline{A} \in (\mathbb{S}_k)^n, k \in \mathbb{N}^* \tag{6.15}$$

which is in turn equivalent to

$$\lambda^d(f) := \sup_{b} \quad b$$
$$\text{s.t.} \quad f(\underline{x}) - b \in \Sigma\langle \underline{x} \rangle_d \tag{6.16}$$

as a consequence of Theorem 6.14.

The dual of SDP (6.16) is

$$\eta^d(f) := \inf_{\mathbf{y}} \quad L_{\mathbf{y}}(f)$$
$$\text{s.t.} \quad y_1 = 1, \quad \mathbf{M}_d(\mathbf{y}) \succeq 0. \tag{6.17}$$

One can compute $\lambda_{\min}(f)$ by solving a single SDP, either SDP (6.17) or SDP (6.16), since there is no duality gap between these two programs, that is, one has $\eta^d(f) = \lambda^d(f) = \lambda_{\min}(f)$.

Now, we address eigenvalue optimization for a given sparse nc polynomial $f = f_1 + \cdots + f_p$ of degree $2d$, with $f_k \in \text{Sym}\,\mathbb{R}\langle \underline{x}, I_k \rangle_{2d}$, for all $k \in [p]$. For all $k \in [p]$, let $\mathbf{G}_{f_k}$ be a Gram matrix associated to f_k. The sparse variant of SDP (6.17) is

$$\eta^d_{cs}(f) := \inf_{\mathbf{y}} \quad L_{\mathbf{y}}(f)$$
$$\text{s.t.} \quad y_1 = 1, \quad \mathbf{M}_d(\mathbf{y}, I_k) \succeq 0, k \in [p] \tag{6.18}$$

whose dual is the sparse variant of SDP (6.16):

$$\lambda^d_{cs}(f) := \sup_{b} \quad b$$
$$\text{s.t.} \quad f - b \in \Sigma\langle \underline{x}, I_1 \rangle_d + \cdots + \Sigma\langle \underline{x}, I_p \rangle_d. \tag{6.19}$$

To prove that there is no duality gap between SDP (6.18) and SDP (6.19), we need the following result.

Proposition 6.16 *The set $\Sigma\langle \underline{x} \rangle^{cs}_d$ is a closed convex subset of $\mathbb{R}\langle \underline{x}, I_1 \rangle_{2d} + \cdots + \mathbb{R}\langle \underline{x}, I_p \rangle_{2d}$.*

From Proposition 6.16, we obtain the following theorem which does not require Assumption 6.4.

Theorem 6.17 *Let $f \in \operatorname{Sym} \mathbb{R}\langle \underline{x}\rangle$ of degree $2d$, with $f = f_1 + \cdots + f_p$, $f_k \in \operatorname{Sym} \mathbb{R}\langle \underline{x}, I_k\rangle_{2d}$, for all $k \in [p]$. Then, one has $\lambda_{cs}^d(f) = \eta_{cs}^d(f)$, i.e., there is no duality gap between SDP (6.18) and SDP (6.19).*

Remark 6.18 *By contrast with the dense case, it is not enough to compute the solution of SDP (6.18) to obtain the optimal value $\lambda_{\min}(f)$ of the unconstrained optimization problem (6.14). However, one can still compute a certified lower bound $\lambda_{cs}^d(f)$ by solving a single SDP, either in the primal form (6.18) or in the dual form (6.19). Note that the related computational cost is potentially much less expensive. Indeed, SDP (6.19) involves PSD matrices of size at most $\max \{\sigma(n_k, d)\}_{k=1}^p$ and $\sum_{k=1}^p \sigma(n_k, 2d)$ equality constraints. This is in contrast with the dense version (6.16), which involves PSD matrices of size $\sigma(n,d)$ and $\sigma(n, 2d)$ equality constraints.*

6.6.2 Constrained eigenvalue optimization

Here, we focus on providing lower bounds for the constrained eigenvalue optimization of nc polynomials. Given $f \in \operatorname{Sym} \mathbb{R}\langle \underline{x}\rangle$ and $\mathfrak{g} = \{g_1, \ldots, g_m\} \subseteq \operatorname{Sym} \mathbb{R}\langle \underline{x}\rangle$ as in (6.2), let us define $\lambda_{\min}(f, \mathfrak{g})$ as follows:

$$\lambda_{\min}(f, \mathfrak{g}) := \inf \{\langle f(\underline{A})\mathbf{v}, \mathbf{v}\rangle : \underline{A} \in \mathcal{D}_{\mathfrak{g}}^\infty, \|\mathbf{v}\| = 1\}, \qquad (6.20)$$

which is, as for the unconstrained case, equivalent to

$$\begin{aligned} \lambda_{\min}(f, \mathfrak{g}) = \ & \sup_b \quad b \\ & \text{s.t.} \quad f(\underline{A}) - b\mathbf{I}_k \succeq 0, \quad \forall \underline{A} \in \mathcal{D}_{\mathfrak{g}}^\infty. \end{aligned} \qquad (6.21)$$

As usual, let $d_j := \lceil \deg(g_j)/2 \rceil$ for each $j \in [m]$, and let

$$r_{\min} := \max \{\lceil \deg(f)/2\rceil, d_1, \ldots, d_m\}.$$

One can approximate $\lambda_{\min}(f, \mathfrak{g})$ from below via the following hierarchy of SDP programs, indexed by $r \geq r_{\min}$:

$$\begin{aligned} \lambda^r(f, \mathfrak{g}) := \ & \sup_b \quad b \\ & \text{s.t.} \quad f - b \in \mathcal{M}(\mathfrak{g})_r \end{aligned} \qquad (6.22)$$

The dual of SDP (6.22) is

$$\begin{aligned} \eta^r(f, \mathfrak{g}) := \ & \inf_{\mathbf{y}} \quad L_{\mathbf{y}}(f) \\ & \text{s.t.} \quad y_1 = 1, \quad \mathbf{M}_r(\mathbf{y}) \succeq 0 \\ & \qquad \mathbf{M}_{r-d_j}(g_j\mathbf{y}) \succeq 0, \quad j \in [m]. \end{aligned} \qquad (6.23)$$

Under additional assumptions, this hierarchy of primal-dual SDP (6.22)–(6.23) converges to the optimal value of the constrained eigenvalue problem.

Theorem 6.19 *Assume that $\mathcal{D}_{\mathfrak{g}}$ is as in (6.6) with the additional quadratic constraints (6.5) and that the quadratic module $\mathcal{M}(\mathfrak{g})$ is Archimedean. Then the following holds for $f \in \operatorname{Sym} \mathbb{R}\langle \underline{x} \rangle$:*

$$\lim_{r \to \infty} \eta^r(f, \mathfrak{g}) = \lim_{r \to \infty} \lambda^r(f, \mathfrak{g}) = \lambda_{\min}(f, \mathfrak{g}). \tag{6.24}$$

The main ingredient of the proof is the nc analog of Putinar's Positivstellensatz, stated in Theorem 6.2.

Let $\mathcal{D}_{\mathfrak{g}}$ be as in (6.6) with the additional quadratic constraints (6.5). Let $\mathcal{M}(\mathfrak{g})^{cs}$ be as in (6.7) and let us define $\mathcal{M}(\mathfrak{g})^{cs}_r$ in the same way as the truncated quadratic module $\mathcal{M}(\mathfrak{g})_r$ in (6.4). Now, let us state the sparse variant of the primal-dual hierarchy (6.22)–(6.23) of lower bounds for $\lambda_{\min}(f, \mathfrak{g})$.

For $r \geq r_{\min}$, the sparse variant of SDP (6.23) is

$$
\begin{aligned}
\eta^r_{cs}(f, \mathfrak{g}) := \quad &\inf_{\mathbf{y}} \quad L_{\mathbf{y}}(f) \\
&\text{s.t.} \quad y_1 = 1, \quad \mathbf{M}_r(\mathbf{y}, I_k) \succeq 0, k \in [p] \\
&\qquad\quad \mathbf{M}_{r-d_j}(g_j \mathbf{y}, I_k) \succeq 0, \quad j \in J_k, k \in [p]
\end{aligned}
\tag{6.25}
$$

whose dual is the sparse variant of SDP (6.22):

$$
\begin{aligned}
\lambda^r_{cs}(f, \mathfrak{g}) := \quad &\sup_{b} \quad b \\
&\text{s.t.} \quad f - b \in \mathcal{M}(\mathfrak{g})^{cs}_r.
\end{aligned}
\tag{6.26}
$$

An ε-neighborhood of 0 is the set $\mathcal{N}_\varepsilon$ defined for a given $\varepsilon > 0$ by

$$\mathcal{N}_\varepsilon := \bigcup_{k \in \mathbb{N}^*} \left\{ (A_1, \ldots, A_n) \in (\mathbb{S}_k)^n : \varepsilon^2 - \sum_{i=1}^{n} A_i^2 \succeq 0 \right\}.$$

Proposition 6.20 *Let $\{f\} \cup \mathfrak{g} \subseteq \operatorname{Sym} \mathbb{R}\langle \underline{x} \rangle$. Assume that $\mathcal{D}_{\mathfrak{g}}$ contains an ε-neighborhood of 0 and that $\mathcal{D}_{\mathfrak{g}}$ is as in (6.6) with the additional quadratic constraints (6.5). Then SDP (6.25) admits strictly feasible solutions. As a result, there is no duality gap between SDP (6.25) and its dual (6.26).*

Moreover, we have the following convergence result implied by Theorem 6.9.

Theorem 6.21 *Let $\{f\} \cup \mathfrak{g} \subseteq \operatorname{Sym} \mathbb{R}\langle \underline{x} \rangle$. Assume that $\mathcal{D}_{\mathfrak{g}}$ is as in (6.6) with the additional quadratic constraints (6.5). Let Assumption 6.4 hold. Then, one has*

$$\lim_{r \to \infty} \eta^r_{cs}(f, \mathfrak{g}) = \lim_{r \to \infty} \lambda^r_{cs}(f, \mathfrak{g}) = \lambda_{\min}(f, \mathfrak{g}). \tag{6.27}$$

As for the unconstrained case, there is no sparse variant of the "perfect" Positivstellensatz, for constrained eigenvalue optimization over convex nc semialgebraic sets [BKP16, Section 4.4], such as those associated either to the sparse nc ball $\mathbb{B}^{cs} := \{1 - \sum_{i \in I_1} x_i^2, \ldots, 1 - \sum_{i \in I_p} x_i^2\}$ or the nc polydisc $\mathbb{D} := \{1 - x_1^2, \ldots, 1 - x_n^2\}$. Namely, for an nc polynomial f of degree $2d + 1$, computing only SDP (6.18) with optimal value $\lambda_{cs}^{d+1}(f, \mathfrak{g})$ when $\mathfrak{g} = \mathbb{B}^{cs}$ or $\mathfrak{g} = \mathbb{D}$ does not suffice to obtain the value of $\lambda_{\min}(f, \mathfrak{g})$. This is explained in Example 6.22 below.

Example 6.22 *Let us consider a randomly generated cubic polynomial $f = f_1 + f_2$ with*

$$
\begin{aligned}
f_1 =\ & 4 - x_1 + 3x_2 - 3x_3 - 3x_1^2 - 7x_1x_2 + 6x_1x_3 - x_2x_1 - 5x_3x_1 + 5x_3x_2 \\
& - 5x_1^3 - 3x_1^2x_3 + 4x_1x_2x_1 - 6x_1x_2x_3 + 7x_1x_3x_1 + 2x_1x_3x_2 - x_1x_3^2 \\
& - x_2x_1^2 + 3x_2x_1x_2 - x_2x_1x_3 - 2x_3^3 - 5x_2^2x_3 - 4x_2x_3^2 - 5x_3x_1^2 \\
& + 7x_3x_1x_2 + 6x_3x_2x_1 - 4x_3x_2x_2 - x_3^2x_1 - 2x_3^2x_2 + 7x_3^3, \\
f_2 =\ & -1 + 6x_2 + 5x_3 + 3x_4 - 5x_2^2 + 2x_2x_3 + 4x_2x_4 - 4x_3x_2 + x_3^2 - x_3x_4 \\
& + x_4x_2 - x_4x_3 + 2x_4^2 - 7x_2^3 + 4x_2x_3^2 + 5x_2x_3x_4 - 7x_2x_4x_3 - 7x_2x_4^2 \\
& + x_3x_2^2 + 6x_3x_2x_3 - 6x_3x_2x_4 - 3x_3^2x_2 - 7x_3^2x_4 + 6x_3x_4x_2 \\
& - 3x_3x_4x_3 - 7x_3x_4^2 + 3x_4x_2^2 - 7x_4x_2x_3 - x_4x_2x_4 - 5x_4x_3^2 \\
& + 7x_4x_3x_4 + 6x_4^2x_2 - 4x_4^3,
\end{aligned}
$$

and the nc polyball $\mathfrak{g} = \mathbb{B}^{cs} = \{1 - x_1^2 - x_2^2 - x_3^2, 1 - x_2^2 - x_3^2 - x_4^2\}$ corresponding to $I_1 = \{1, 2, 3\}$ and $I_2 = \{2, 3, 4\}$. Then, one has $\lambda_{cs}^2(f, \mathfrak{g}) \simeq -27.536 < \lambda_{cs}^3(f, \mathfrak{g}) \simeq -27.467 \simeq \lambda_{\min}^2(f, \mathfrak{g}) = \lambda_{\min}(f, \mathfrak{g})$. In Appendix B.2, we provide a Julia script to compute these bounds.

6.6.3 Extracting optimizers

Here, we explain how to extract a pair of optimizers $(\underline{A}, \mathbf{v})$ for the eigenvalue optimization problems when the flatness and irreducibility conditions of Theorem 6.12 hold. We apply the SparseGNS procedure on the optimal solution of SDP (6.18) in the unconstrained case or SDP (6.25) in the constrained case.

Proposition 6.23 *Given f as in Theorem 6.17, let us assume that SDP (6.18) (with d being replaced by $d + 1$) yields an optimal solution $\mathbf{y}$ associated to $\eta_{cs}^{d+1}(f)$. If the sequence $\mathbf{y}$ satisfies the flatness (H1) and irreducibility (H2) conditions stated in Theorem 6.12, then one has*

$$
\lambda_{\min}(f) = \eta_{cs}^{d+1}(f) = L_{\mathbf{y}}(f).
$$

Algorithm 3 `SparseEigGNS`

Require: $f \in \operatorname{Sym} \mathbb{R}\langle \underline{x} \rangle_{2d}$ satisfying Assumption 6.4
Ensure: $\underline{A}$ and $\mathbf{v}$
 1: Compute $\eta_{cs}^{d+1}(f)$ by solving SDP (6.18)
 2: **if** SDP (6.18) is unbounded or its optimum is not attained **then**
 3:     Stop
 4: **end if**
 5: Let $\mathbf{M}_{d+1}(\mathbf{y})$ be an optimizer of SDP (6.18)
 6: Compute $\underline{A}, \mathbf{v} := \mathtt{SparseGNS}(\mathbf{M}_{d+1}(\mathbf{y}))$

We can extract optimizers for the unconstrained minimal eigenvalue problem (6.14) thanks to the following algorithm.

In the constrained case, the next result is a direct corollary of Theorem 6.12.

Corollary 6.24 *Let* $\{f\} \cup \mathfrak{g} \subseteq \operatorname{Sym} \mathbb{R}\langle \underline{x} \rangle$, *and assume that* $\mathcal{D}_{\mathfrak{g}}$ *is as in (6.6) with the additional quadratic constraints (6.5). Suppose Assumptions 6.4(i)–(ii) hold. Let* $\mathbf{y}$ *be an optimal solution of SDP (6.25) with optimal value* $\eta_{cs}^r(f, \mathfrak{g})$ *for* $r \geq r_{\min} + \delta$, *such that* $\mathbf{y}$ *satisfies the assumptions of Theorem 6.12. Then, there exist* $t \in \mathbb{N}^*, \underline{A} \in \mathcal{D}_{\mathfrak{g}}^t$ *and a unit vector* $\mathbf{v}$ *such that*

$$\lambda_{\min}(f, \mathfrak{g}) = \langle f(\underline{A})\mathbf{v}, \mathbf{v} \rangle = \eta_{cs}^r(f, \mathfrak{g}).$$

Example 6.25 *Consider the sparse polynomial* $f = f_1 + f_2$ *from Example 6.22. The moment matrix* $\mathbf{M}_3(\mathbf{y})$ *obtained by solving (6.25) with* $r = 3$ *satisfies the flatness (H1) and irreducibility (H2) conditions of Theorem 6.12. We can thus apply the* SparseGNS *algorithm yielding*

$$A_1 = \begin{bmatrix} 0.0059 & 0.0481 & 0.1638 & 0.4570 \\ 0.0481 & -0.2583 & 0.5629 & -0.2624 \\ 0.1638 & 0.5629 & 0.3265 & -0.3734 \\ 0.4570 & -0.2624 & -0.3734 & -0.2337 \end{bmatrix}$$

$$A_2 = \begin{bmatrix} -0.3502 & 0.0080 & 0.1411 & 0.0865 \\ 0.0080 & -0.4053 & 0.2404 & -0.1649 \\ 0.1411 & 0.2404 & -0.0959 & 0.3652 \\ 0.0865 & -0.1649 & 0.3652 & 0.4117 \end{bmatrix}$$

$$A_3 = \begin{bmatrix} -0.7669 & -0.0074 & -0.1313 & -0.0805 \\ -0.0074 & -0.4715 & -0.2238 & 0.1535 \\ -0.1313 & -0.2238 & 0.0848 & -0.3400 \\ -0.0805 & 0.1535 & -0.3400 & -0.2126 \end{bmatrix}$$

$$A_4 = \begin{bmatrix} 0.3302 & -0.1839 & 0.1811 & -0.0404 \\ -0.1839 & -0.1069 & 0.5114 & -0.0570 \\ 0.1811 & 0.5114 & 0.1311 & -0.3664 \\ -0.0404 & -0.0570 & -0.3664 & 0.4440 \end{bmatrix}$$

where

$$f(\underline{A}) = \begin{bmatrix} -10.3144 & 3.9233 & -5.0836 & -7.7828 \\ 3.9233 & 1.8363 & 4.5078 & -7.5905 \\ -5.0836 & 4.5078 & -19.5827 & 13.9157 \\ -7.7828 & -7.5905 & 13.9157 & 8.3381 \end{bmatrix}$$

has minimal eigenvalue -27.4665 *with unit eigenvector*

$$\mathbf{v} = \begin{bmatrix} 0.1546 & -0.2507 & 0.8840 & -0.3631 \end{bmatrix}^{\mathsf{T}}.$$

In this case all the ranks involved are equal to four. So A_2 and A_3 are computed from $\mathbf{M}_3(\mathbf{y}, I_1 \cap I_2)$, after an appropriate basis change A_1 (and the same A_2, A_3) is obtained from $\mathbf{M}_3(\mathbf{y}, I_1)$, and finally A_4 is computed from $\mathbf{M}_3(\mathbf{y}, I_2)$.

6.7 Overview of numerical experiments

The aim of this section is to provide experimental comparison between the bounds given by the dense hierarchy and the ones produced by our CS variant. The numerical results were obtained with the Julia package NCTSSOS employing Mosek as an SDP solver. The computation was carried out on Intel(R) Core(TM) i9-10900 CPU@2.80GHz with 64G RAM.

6.7.1 An unconstrained problem

In Table 6.1, we report results obtained for minimizing the eigenvalue of the nc variant of the chained singular function [CGT88]:

$$f = \sum_{i \in J} \left((x_i + 10x_{i+1})^2 + 5(x_{i+2} - x_{i+3})^2 \right.$$

$$\left. + (x_{i+1} - 2x_{i+2})^4 + 10(x_i - 10x_{i+3})^4 \right),$$

where $J = [n-3]$ and n is a multiple of 4. We compute lower bounds on the minimal eigenvalue of f for $n = 40, 80, 120, 160, 200, 240$. For each value of n, "mb" stands for maximal sizes of PSD blocks involved either in the sparse relaxation (6.18) or the dense relaxation (6.17). As one can see, the size of the SDP programs is significantly reduced after exploiting CS, which is consistent with Remark 6.18. In addition, the sparse approach turns out to be much more efficient and scalable than the dense approach.

6.7.2 Bell inequalities

Upper bounds on quantum violations of Bell inequalities can be computed using eigenvalue maximization of nc polynomials. The classical (also most concise) Bell inequality states that $v^\star (A_1 \otimes B_1 + A_1 \otimes B_2 + A_2 \otimes$

Table 6.1: Sparse versus dense approaches for minimizing eigenvalues of the chained singular function. mb: maximal size of PSD blocks, opt: optimum, time: running time in seconds. "-" indicates an out of memory error.

n	sparse			dense		
	mb	opt	time	mb	opt	time
40	13	0	0.17	157	0	142
80	13	0	0.43	-	-	-
120	13	0	0.65	-	-	-
160	13	0	0.89	-	-	-
200	13	-0.0014	1.02	-	-	-
240	13	-0.0016	1.28	-	-	-

$B_1 - A_2 \otimes B_2)v$ is at most 2 for all separable states $v \in \mathbb{C}^k \otimes \mathbb{C}^k$ and self-adjoint $A_j, B_j \in \mathbb{C}^{k \times k}$ with $A_j^2 = B_j^2 = \mathbf{I}_k$. The so-called Tsirelson's bound implies that the above quantity is at most $2\sqrt{2}$ when one allows arbitrary states. This bound on the maximum violation level can be obtained by eigenvalue-maximizing $a_1 b_1 + a_1 b_2 + a_2 b_1 - a_2 b_2$ under the constraints $a_j^2 = b_j^2 = 1$ and $a_i b_j = b_j a_i$. To show the potential benefits of our approach based on CS, we consider the Bell inequality, called I_{3322}, and compute upper bounds of its maximum violation level. The associated objective function is $f = a_1(b_1 + b_2 + b_3) + a_2(b_1 + b_2 - b_3) + a_3(b_1 - b_2) - a_1 - 2b_1 - b_2$. The set of constraints is $a_j^2 = a_j$, $b_j^2 = b_j$, and $a_i b_j = b_j a_i$. Table 6.2 compares the efficiency and accuracy of the sparse approach with the dense one, for different relaxation orders. It can be seen that the sparse approach spends much less time while providing almost the same bounds.

Table 6.2: Sparse versus dense approaches for maximizing the violation level of the Bell inequality I_{3322}. r denotes the relaxation order.

r	sparse			dense		
	mb	opt	time	mb	opt	time
2	28	0.2509398	0.01	13	0.2590718	0.01
3	88	0.2508758	0.22	25	0.2512781	0.02
4	244	0.2508754	8.40	41	0.2509057	0.02
5	628	0.2508752	456	61	0.2508774	0.04
6	-	-	-	85	0.2508754	0.09

6.8   Notes and sources

The main results presented in this chapter have been published in [KMP21]. Applications of interest related to noncommutative optimization arise from quantum theory and quantum information science [NPA08, PKRR+19] as well as control theory [SIG98, dOHMP09]. Further motivation relates to the generalized Lax conjecture [Lax58], where the goal is to get computer-assisted proofs based on SOHS in Clifford algebras [NT14]. The verification of noncommutative polynomial trace inequalities has also been motivated by a conjecture formulated by Bessis, Moussa and Villani (BMV) in 1975 [BMV75], which has been recently proved by Stahl [Sta13] (see also the Lieb and Seiringer reformulation [LS04]). Further efforts focused on applications arising from bipartite quantum correlations [GdLL18], and matrix factorization ranks in [GDLL19]. In a related analytic direction, there has been recent progress dedicated to multivariate generalizations of the Golden-Thompson inequality and the Araki-Lieb-Thirring inequality [SBT17, HKT17].

There is a plethora of prior research in quantum information theory involving reformulating problems as optimization of noncommutative polynomials. One famous application is to characterize the set of quantum correlations. Bell inequalities [Bel64] provide a method to investigate entanglement, which allows two or more parties to be correlated in a nonclassical way, and is often studied through the set of bipartite quantum correlations. Such correlations consist of the conditional probabilities that two physically separated parties can generate by performing measurements on a shared entangled state. These conditional probabilities satisfy some inequalities classically, but violate them in the quantum realm [CHSH69].

In this context, a given noncommutative polynomial in n variables and of degree $2d$ is PSD if and only if it decomposes as a SOHS [Hel02, McC01]. In practice, an SOHS decomposition can be computed by solving an SDP involving PSD matrices of size $O(n^d)$, which is even larger than the size of the matrices involved in the commutative case. SOHS decompositions are also used for constrained optimization, either to minimize eigenvalues or traces of noncommutative polynomial objective functions, under noncommutative polynomial (in)equality constraints. The optimal value of such constrained problems can be approximated, as closely as desired, while relying on the noncommutative analogue of Lasserre's hierarchy [PNA10, CKP12, BCKP13]. The NCSOStools [CKP11, BKP16] library can compute such approximations for optimization problems involving polynomials in noncommuting variables. By comparison with the commutative case, the size $O(n^r)$ of the SDP matrices at a given step r of the noncommutative hierarchy becomes intractable even faster.

A remedy for unconstrained problems is to rely on the adequate non-commutative analogue of the standard Newton polytope method, which is called the *Newton chip method* (see e.g., [BKP16, §2.3]) and can be further improved with the *augmented Newton chip method* (see e.g., [BKP16, §2.4]), by removing certain terms which can never appear in an SOHS decomposition of a given input. As in the commutative case, the Newton polytope method cannot be applied for constrained problems. When one cannot go from step r to step $r + 1$ in the hierarchy because of the computational burden, one can always consider matrices indexed by all terms of degree r plus a fixed percentage of terms of degree $r + 1$. This is used for instance to compute tighter upper bounds for maximum violation levels of Bell inequalities [PV09]. Another trick, implemented in the Ncpol2sdpa library [Wit15], consists of exploiting simple equality constraints, such as "$x^2 = 1$", to derive substitution rules for variables involved in the SDP relaxations. Similar substitutions are performed in the commutative case by GloptiPoly [HLL09].

Proposition 6.1 can be found, e.g., in [Hel02, §2.2]. The noncommutative analog of Putinar's Positivstellensatz is due to Helton and McCullough [HM04, Theorem 1.2]. Lemma 6.5 is proved in [BKP16, Lemma 1.44]. Theorem 6.8 is proved in [BKP16, Theorem 1.27]. For more details on amalgamation theory for C^*-algebras, see, e.g., [Bla78, Voi85]. Theorem 6.7 can be found in [Bla78] or [Voi85, Section 5]. This amalgamation theory serves to prove our sparse representation result 6.9, originally stated in [KMP21, Theorem 3.3].

The notion of flatness was used in a noncommutative setting for the first time by McCullough [McC01] in his proof of the Helton-McCullough theorem, cf. [McC01, Lemma 2.2]. In the dense case [PNA10] (see also [AL12, Chapter 21] and [BKP16, Theorem 1.69]) provides a first noncommutative variant for the eigenvalue problem. See [BCKP13] for a similar construction for the trace problem. The sparse version of the flat extension theorem is stated in [KMP21, Theorem 4.2] and the SparseGNS algorithm is explicitly given in [KMP21, Algorithm 4.6] for the case of two subsets of variables (the general case is similar). Theorem 6.12 can be seen as a noncommutative variant of the result by Lasserre stated in [Las06, Theorem 3.7], related to the minimizer extraction in the context of sparse polynomial optimization. In the sparse commutative case, Lasserre assumes flatness of each moment matrix indexed by the canonical basis of $\mathbb{R}[\mathbf{x}, I_k]_r$, for each $k \in [p]$, which is similar to our flatness condition (H1). The difference is that this technical flatness condition on each I_k adapts to the degree of the constraint polynomials in variables in I_k, resulting in an adapted parameter δ_k instead of global δ. We could assume the same in Theorem 6.12 but for the sake of simplicity, we assume that these parameters are all equal. In addition, Lasserre assumes that each moment matrix indexed by the canonical basis of $\mathbb{R}[\underline{x}, I_j \cap I_k)]_r$ is of rank one, for all pairs (j, k) with $I_j \cap I_k \neq \varnothing$, which is the commutative analog of our irreducibility condition (H2).

The absence of duality gap for unconstrained eigenvalue minimization is derived in the dense case, e.g., in [BKP16, Theorem 4.1] and relies on [MP05, Proposition 3.4], which says that the set of SOHS polynomials is closed in the set of nc polynomials. Proposition 6.16 is a sparse version of this latter result, leading to Theorem 6.17, originally proved in [KMP21, Theorem 5.4].

The "perfect" Positivstellensatz for constrained eigenvalue optimization over convex nc semialgebraic set (e.g., the nc ball or the nc polydisc) is stated in [BKP16, §4.4] or [HKM12]. Proposition 6.23 and Corollary 6.24 are the sparse variants of [BKP16, Proposition 4.4] and [BKP16, Theorem 4.12], in the unconstrained and constrained settings, respectively.

As in the dense case [BKP16, Algorithm 4.2], one can provide a randomized algorithm to look for flat optimal solutions for the constrained eigenvalue problem (6.20). The underlying reason which motivates this randomized approach is work by Nie, who derives in [Nie14] a hierarchy of SDP programs, with a random objective function, that converges to a flat solution (under mild assumptions).

The interested reader can find more about trace minimization of sparse polynomials in [KMP21, §6] and more detailed numerical experiments in [KMP21, §7]. For a detailed account about maximal violation levels of Bell inequalities (in particular the one mentioned in Table 6.2), we refer to [PV09]. CS can be exploited in a similar way to solve trace polynomial optimization problems [KMV22].

Bibliography

[AL12]   Miguel F. Anjos and Jean-Bernard Lasserre, editors. *Handbook on semidefinite, conic and polynomial optimization*, volume 166 of *International Series in Operations Research & Management Science*. Springer, New York, 2012.

[BCKP13]   Sabine Burgdorf, Kristijan Cafuta, Igor Klep, and Janez Povh. The tracial moment problem and trace-optimization of polynomials. *Math. Program.*, 137(1-2, Ser. A):557–578, 2013.

[Bel64]   John S Bell. On the Einstein Podolsky Rosen paradox. *Physics Physique Fizika*, 1(3):195, 1964.

[BKP16]   Sabine Burgdorf, Igor Klep, and Janez Povh. *Optimization of polynomials in non-commuting variables*. SpringerBriefs in Mathematics. Springer, [Cham], 2016.

[Bla78]   Bruce E. Blackadar. Weak expectations and nuclear C^*-algebras. *Indiana Univ. Math. J.*, 27(6):1021–1026, 1978.

[BMV75]   Daniel Bessis, Pierre Moussa, and Matteo Villani. Monotonic converging variational approximations to the functional integrals in quantum statistical mechanics. *J. Math. Phys.*, 16(11):2318–2325, 1975.

[CGT88]   Andrew R. Conn, Nicholas I. M. Gould, and Philippe L. Toint. Testing a class of methods for solving minimization problems with simple bounds on the variables. *Math. Comp.*, 50(182):399–430, 1988.

[CHSH69]   John F. Clauser, Michael A. Horne, Abner Shimony, and Richard A. Holt. Proposed experiment to test local hidden-variable theories. *Phys. rev. lett.*, 23(15):880, 1969.

[CKP11]   Kristijan Cafuta, Igor Klep, and Janez Povh. NCSOStools: a computer algebra system for symbolic and numerical computation with noncommutative polynomials. *Optim. Methods Softw.*, 26(3):363–380, 2011.

[CKP12]   Kristijan Cafuta, Igor Klep, and Janez Povh. Constrained
           polynomial optimization problems with noncommuting vari-
           ables. *SIAM J. Optim.*, 22(2):363–383, 2012.

[dOHMP09] Mauricio C. de Oliveira, J. William Helton, Scott A. Mc-
           Cullough, and Mihai Putinar. Engineering systems and free
           semi-algebraic geometry. In *Emerging applications of algebraic
           geometry*, volume 149 of *IMA Vol. Math. Appl.*, pages 17–61.
           Springer, New York, 2009.

[GdLL18]   Sander Gribling, David de Laat, and Monique Laurent.
           Bounds on entanglement dimensions and quantum graph
           parameters via noncommutative polynomial optimization.
           Math. Program., 170(1, Ser. B):5–42, 2018.

[GDLL19]   Sander Gribling, David De Laat, and Monique Laurent.
           Lower bounds on matrix factorization ranks via noncommu-
           tative polynomial optimization. *Foundations of Computational
           Mathematics*, 19(5):1013–1070, 2019.

[Hel02]   J. William Helton. "Positive" noncommutative polynomials
           are sums of squares. *Ann. of Math. (2)*, 156(2):675–694, 2002.

[HKM12]   J. William Helton, Igor Klep, and Scott McCullough. The con-
           vex Positivstellensatz in a free algebra. *Adv. Math.*, 231(1):516–
           534, 2012.

[HKT17]   Fumio Hiai, Robert König, and Marco Tomamichel. Gen-
           eralized log-majorization and multivariate trace inequalities.
           Ann. Henri Poincaré, 18(7):2499–2521, 2017.

[HLL09]   D. Henrion, Jean-Bernard Lasserre, and J. Löfberg. Glop-
           tiPoly 3: moments, optimization and semidefinite program-
           ming. *Optimization Methods and Software*, 24(4-5):pp. 761–779,
           August 2009.

[HM04]   J. William Helton and Scott A. McCullough. A Positivstel-
           lensatz for non-commutative polynomials. *Trans. Amer. Math.
           Soc.*, 356(9):3721–3737, 2004.

[KMP21]   Igor Klep, Victor Magron, and Janez Povh. Sparse noncom-
           mutative polynomial optimization. *Mathematical Program-
           ming*, pages 1–41, 2021.

[KMV22]   Igor Klep, Victor Magron, and Jurij Volčič. Optimization
           over trace polynomials. In *Annales Henri Poincaré*, volume 23,
           pages 67–100. Springer, 2022.

[Las06]   Jean B Lasserre. Convergent sdp-relaxations in polynomial optimization with sparsity. *SIAM Journal on Optimization*, 17(3):822–843, 2006.

[Lax58]   Peter D. Lax. Differential equations, difference equations and matrix theory. *Comm. Pure Appl. Math.*, 11:175–194, 1958.

[LS04]   Elliott H. Lieb and Robert Seiringer. Equivalent forms of the Bessis-Moussa-Villani conjecture. *J. Statist. Phys.*, 115(1-2):185–190, 2004.

[McC01]   Scott McCullough. Factorization of operator-valued polynomials in several non-commuting variables. *Linear Algebra Appl.*, 326(1-3):193–203, 2001.

[MP05]   Scott McCullough and Mihai Putinar. Noncommutative sums of squares. *Pacific J. Math.*, 218(1):167–171, 2005.

[Nie14]   Jiawang Nie. The *A*-truncated *K*-moment problem. *Found. Comput. Math.*, 14(6):1243–1276, 2014.

[NPA08]   Miguel Navascués, Stefano Pironio, and Antonio Acín. A convergent hierarchy of semidefinite programs characterizing the set of quantum correlations. *New J. Phys.*, 10(7):073013, 2008.

[NT14]   Tim Netzer and Andreas Thom. Hyperbolic polynomials and generalized Clifford algebras. *Discrete Comput. Geom.*, 51(4):802–814, 2014.

[PKRR+19]   Alejandro Pozas-Kerstjens, Rafael Rabelo, Łukasz Rudnicki, Rafael Chaves, Daniel Cavalcanti, Miguel Navascués, and Antonio Acín. Bounding the sets of classical and quantum correlations in networks. *Phys. Rev. Lett.*, 123(14):140503, 6, 2019.

[PNA10]   Stefano Pironio, Miguel Navascués, and Antonio Acín. Convergent relaxations of polynomial optimization problems with noncommuting variables. *SIAM J. Optim.*, 20(5):2157–2180, 2010.

[PV09]   Károly F. Pál and Tamás Vértesi. Quantum bounds on Bell inequalities. *Phys. Rev. A (3)*, 79(2):022120, 12, 2009.

[SBT17]   David Sutter, Mario Berta, and Marco Tomamichel. Multivariate trace inequalities. *Comm. Math. Phys.*, 352(1):37–58, 2017.

[SIG98]   Robert E. Skelton, Tetsuya Iwasaki, and Karolos M. Grigoriadis. *A unified algebraic approach to linear control design*. The Taylor & Francis Systems and Control Book Series. Taylor & Francis, Ltd., London, 1998.

[Sta13]   Herbert R. Stahl. Proof of the BMV conjecture. *Acta Math.*,
            211(2):255–290, 2013.

[Voi85]   Dan-Virgil Voiculescu. Symmetries of some reduced free
            product C^*-algebras. In *Operator algebras and their connections
            with topology and ergodic theory (Busteni, 1983)*, volume 1132 of
            Lecture Notes in Math., pages 556–588. Springer, Berlin, 1985.

[Wit15]   Peter Wittek. Algorithm 950: Ncpol2sdpa-sparse semidefinite
            programming relaxations for polynomial optimization prob-
            lems of noncommuting variables. *ACM Trans. Math. Software*,
            41(3):Art. 21, 12, 2015.

Part III

Term sparsity

Chapter 7

The moment-SOS hierarchy based on term sparsity

As emphasized earlier for distinct applications, exploiting CS arising from a POP may allow to significantly reduce the computational cost of the related hierarchy of SDP relaxations assuming that the csp is sufficiently sparse. Nevertheless a POP can be fairly sparse (namely, involving only a small number of terms) whereas its csp is (nearly) dense. For instance, if some constraint (e.g., $1 - \|\mathbf{x}\|_2^2 \geq 0$) involves all variables, then the csp is dense. On the other hand, instead of exploiting sparsity from the perspective of *variables*, one can also exploit sparsity from the perspective of *terms*, which leads to the notions of *term sparsity (TS)* and *term sparsity pattern (tsp)*.

Roughly speaking, the tsp can be also represented by a graph, which is called the tsp graph. But unlike the csp graph, the nodes of the tsp graph are monomials (coming from a monomial basis) and the edges of the graph grasp the links between monomials emerging from the related SOS decomposition. We are able to design an iterative procedure to enlarge the tsp graph in order to iteratively exploit TS for the given POP. Each iteration consists of two successive operations: (i) a support extension followed by (ii) a chordal extension. In doing so we obtain a finite ascending chain of graphs:

$$G^{(1)} \subseteq G^{(2)} \subseteq \cdots \subseteq G^{(s)} = G^{(s+1)}.$$

Then combining this iterative procedure with the standard moment-SOS hierarchy results in a two-level moment-SOS hierarchy involving multiple PSD blocks. When the sizes of the blocks are small, the associated SDP relaxations can be drastically much cheaper to solve.

To some extent, TS (working on the monomial level) is finer than CS (working on the variable level) in describing sparsity for a POP. A natural idea for solving large-scale POPs then is: first exploiting CS to decompose variables into a set of cliques, and second exploiting TS for each subsystem involving only one variable clique to further reduce the size of SDPs. This idea has been successfully carried out in [WMLM20] and will be presented later on, in Chapter 8.

Section 7.1 focuses on exploiting TS for unconstrained POPs. We prove in Section 7.2 that the resulting scheme is always more accurate than the framework based on *scaled diagonally dominant* sum of squares (SDSOS). Section 7.3 presents a TS variant of the moment-SOS hierarchy of SDP relaxations for general POPs with compact constraints. Next, we provide in Section 7.4 more sophisticated algorithms to reduce even further the size of the resulting SDP relaxations. Section 7.5 explains the relation between the block structures arising from the TS-adapted relaxations and the one provided by sign symmetries. Eventually, we illustrate in Section 7.6 the efficiency and the accuracy of the TS-adapted relaxations on benchmarks coming from the global optimization literature and networked systems stability.

7.1   The TSSOS hierarchy for unconstrained POP

In this section, we describe an iterative procedure to exploit TS for the primal-dual moment-SOS relaxations of unconstrained POPs. Recall the formulation of an unconstrained POP:

$$\mathbf{P}: \quad f_{\min} := \inf \{f(\mathbf{x}) : \mathbf{x} \in \mathbb{R}^n\} = \sup \{b : f - b \geq 0\}, \qquad (7.1)$$

where $f = \sum_\alpha f_\alpha \mathbf{x}^\alpha \in \mathbb{R}[\mathbf{x}]$. Suppose that f is of degree $2d$ with $\text{supp}(f) = \mathscr{A}$ (w.l.o.g. assuming $\mathbf{0} \in \mathscr{A}$) and $\mathbf{x}^{\mathscr{B}} := (\mathbf{x}^\beta)_{\beta \in \mathscr{B}}$ is a monomial basis arranged with respect to any fixed ordering. For convenience, we slightly abuse notation in the sequel and denote by $\mathscr{B}$ (resp. β) instead of $\mathbf{x}^{\mathscr{B}}$ (resp. $\mathbf{x}^\beta$) a monomial basis (resp. a monomial). One may choose $\mathscr{B}$ to be the standard monomial basis $\mathbb{N}_d^n$. But when f is sparse, the following theorem due to Reznick [Rez78] allows us to use a (possibly) smaller monomial basis by considering Newton polytopes. Recall that for a polynomial $f \in \mathbb{R}[\mathbf{x}]$, the *Newton polytope* of f, denoted by $\mathcal{N}(f)$, is the convex hull generated by its support.

Theorem 7.1 *If $f, p_i \in \mathbb{R}[\mathbf{x}]$ and $f = \sum_i p_i^2$, then $\mathcal{N}(p_i) \subseteq \frac{1}{2}\mathcal{N}(f)$.*

As an immediate corollary, we can take the integer points in half of the Newton polytope of f to form a monomial basis, i.e.,

$$\mathscr{B} = \frac{1}{2}\mathcal{N}(f) \cap \mathbb{N}^n \subseteq \mathbb{N}_d^n. \qquad (7.2)$$

Example 7.2 *Let $f = 4x_1^4 x_2^6 + x_1^2 - x_1 x_2^2 + x_2^2$. Then we have*

$$\text{supp}(f) = \{(4,6),(2,0),(1,2),(0,2)\}$$

and $\mathscr{B} = \frac{1}{2}\mathcal{N}(f) \cap \mathbb{N}^n = \{(1,0),(2,3),(0,1),(1,2),(1,1)\}$ (Figure 7.1).

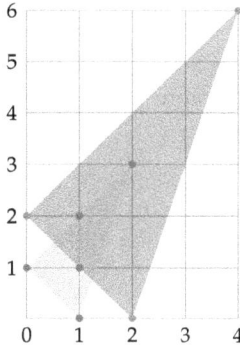

Figure 7.1: The monomial basis given by half of the Newton polytope (marked by red).

Given a monomial basis $\mathscr{B}$ and a sequence $\mathbf{y} \subseteq \mathbb{R}$, the moment matrix $\mathbf{M}_{\mathscr{B}}(\mathbf{y})$ associated with $\mathscr{B}$ and $\mathbf{y}$ is the block of the moment matrix $\mathbf{M}_d(\mathbf{y})$ indexed by $\mathscr{B}$. Then the moment relaxation of $\mathbf{P}$ in the monomial basis $\mathscr{B}$ is given by

$$
\mathbf{P}_{\text{mom}} : \quad
\begin{aligned}
f_{\text{mom}} &:= \inf_{\mathbf{y}} \quad L_{\mathbf{y}}(f) \\
&\text{s.t.} \quad \mathbf{M}_{\mathscr{B}}(\mathbf{y}) \succeq 0 \\
&\qquad\; y_0 = 1.
\end{aligned}
\tag{7.3}
$$

For a graph $G(V, E)$ with $V \subseteq \mathbb{N}^n$, let the support of G be given by

$$\text{supp}(G) := \{\beta + \gamma \mid \beta = \gamma \text{ or } \{\beta, \gamma\} \in E\}. \tag{7.4}$$

We define G^{tsp} to be the graph with nodes $V = \mathscr{B}$ and with edges

$$E(G^{\text{tsp}}) = \{\{\beta, \gamma\} \mid \beta \neq \gamma \in V, \ \beta + \gamma \in \mathscr{A} \cup (2\mathscr{B})\}, \tag{7.5}$$

which is called the tsp graph associated with f.

Starting with the initial graph $G^{(0)} = G^{\text{tsp}}$, we now define a sequence of graphs $(G^{(s)})_{s \geq 1}$ by iteratively performing two successive operations:
(1) **support extension**. Let $F^{(s)}$ be the graph with nodes V and with edges

$$E(F^{(s)}) = \left\{ \{\beta, \gamma\} \mid \beta \neq \gamma \in V, \ \beta + \gamma \in \text{supp}(G^{(s-1)}) \right\}. \tag{7.6}$$

(2) **chordal extension**. Let $G^{(s)} = \left(F^{(s)} \right)'$.

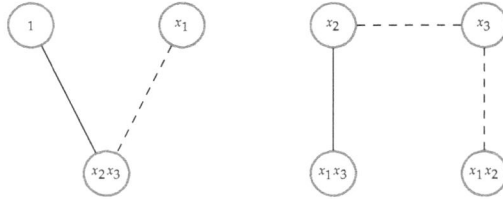

Figure 7.2: The support extension of G in Example 7.3. The dashed edges are added after support extension.

Example 7.3 *Let us consider the graph $G(V, E)$ with*

$$V = \{1, x_1, x_2, x_3, x_2x_3, x_1x_3, x_1x_2\} \text{ and } E = \{\{1, x_2x_3\}, \{x_2, x_1x_3\}\}.$$

Figure 7.2 illustrates the support extension of G.

By construction, one has $G^{(s)} \subseteq G^{(s+1)}$ for $s \geq 1$ and therefore the sequence of graphs $(G^{(s)})_{s\geq 1}$ stabilizes after a finite number of steps. Following what we introduced in Section 1.2, we denote by $\Pi_{G^{(s)}}(S^+_{|\mathscr{B}|})$ the set of matrices in $S(G^{(s)})$ that have a PSD completion, and denote by $\mathbf{B}_{G^{(s)}}$ the adjacency matrix of $G^{(s)}$. If f is sparse, by replacing $\mathbf{M}_{\mathscr{B}}(\mathbf{y}) \succeq 0$ with the weaker condition $\mathbf{M}_{\mathscr{B}}(\mathbf{y}) \in \Pi_{G^{(s)}}(S^+_{|\mathscr{B}|})$ in (7.3), we then obtain a sparse moment relaxation of (7.1) for each $s \geq 1$:

$$\mathbf{P}^s_{ts} : \quad \begin{cases} \underset{\mathbf{y}}{\inf} & L_{\mathbf{y}}(f) \\ \text{s.t.} & \mathbf{B}_{G^{(s)}} \circ \mathbf{M}_{\mathscr{B}}(\mathbf{y}) \in \Pi_{G^{(s)}}(S^+_{|\mathscr{B}|}) \\ & y_0 = 1 \end{cases} \qquad (7.7)$$

with optimum denoted by f^s_{ts}. We call $(\mathbf{P}^s_{ts})_{s\geq 1}$ the TS-adpated moment-SOS (TSSOS) hierarchy for $\mathbf{P}$ and call s the *sparse order*.

Remark 7.4 *The intuition behind the support extension operation is that once one position related to y_α in the moment matrix $\mathbf{M}_{\mathscr{B}}(\mathbf{y})$ is "activated" in the sparsity pattern, then all positions related to y_α in $\mathbf{M}_{\mathscr{B}}(\mathbf{y})$ should be "activated". In addition, Theorems 1.4 and 1.5 provide the rationale behind the mechanism of the chordal extension operation.*

Theorem 7.5 *The sequence $(f^s_{ts})_{s\geq 1}$ is monotonically nondecreasing and $f^s_{ts} \leq f_{mom}$ for all $s \geq 1$.*

PROOF The inclusion $G^{(k)} \subseteq G^{(k+1)}$ implies that each maximal clique of $G^{(k)}$ is a subset of some maximal clique of $G^{(k+1)}$. Thus by Theorem 1.5, we see that $\mathbf{P}^s_{ts}$ is a relaxation of $\mathbf{P}^{s+1}_{ts}$ (and also a relaxation of $\mathbf{P}_{mom}$). This yields the desired conclusions.   □

As a consequence of Theorem 7.5, we obtain the following hierarchy of lower bounds for the optimum of **P**:

$$f_{ts}^1 \leq f_{ts}^2 \leq \cdots \leq f_{mom} \leq f_{min}. \tag{7.8}$$

If the maximal chordal extension is chosen for the chordal extension operation, then we can show (see [WML21] for more details) that the sequence $(f_{ts}^s)_{s \geq 1}$ converges to the global optimum f_{min}. Otherwise, there is no guarantee of such convergence as illustrated by the following example.

Example 7.6 *Consider the commutative version of the polynomial from (6.13):*

$$f = x_1^2 - 2x_1x_2 + 3x_2^2 - 2x_1^2x_2 + 2x_1^2x_2^2 - 2x_2x_3 + 6x_3^2$$
$$+ 18x_2^2x_3 - 54x_2x_3^2 + 142x_2^2x_3^2.$$

The monomial basis computed from the Newton polytope is $\{1, x_1, x_2, x_3, x_1x_2, x_2x_3\}$. Figure 7.3 shows the tsp graph G^{tsp} (without dashed edges) and its smallest chordal extension $G^{(1)}$ (with dashed edges) for f. The graph sequence $(G^{(s)})_{s \geq 1}$ stabilizes at $s = 1$. Solving $\mathbf{P}_{ts}^1$, we obtain $f_{ts}^1 \approx -0.00355$ while $f_{mom} = f_{min} = 0$. On the other hand, note that G^{tsp} has only one connected component. So with the maximal chordal extension, we immediately get the complete graph and it follows $f_{ts}^1 = f_{mom} = 0$ in this case.

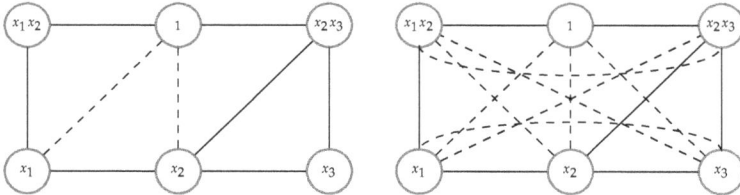

Figure 7.3: The tsp graph G^{tsp} and a smallest chordal extension (left) as well as the maximal chordal extension (right) for Example 7.6.

For each $s \geq 1$, the dual SDP of (7.7) is

$$\begin{cases} \sup_{G,b} & b \\ \text{s.t.} & \langle G, B_\alpha \rangle = f_\alpha - b1_{\alpha=0}, \quad \forall \alpha \in \text{supp}(G^{(s)}) \\ & G \in S_{|\mathcal{B}|}^+ \cap S(G^{(s)}) \end{cases} \tag{7.9}$$

where B_α has been defined after (2.8).

Proposition 7.7 *For each $s \geq 1$, there is no duality gap between $\mathbf{P}_{ts}^s$ and its dual (7.9).*

PROOF This easily follows from the fact that $\mathbf{P}_{ts}^s$ satisfies Slater's condition by Proposition 3.1 of [Las01] and Theorem 1.5.   □

7.2   Comparison with SDSOS

SDSOS polynomials were introduced and studied in [AM14] as cheaper alternatives to SOS polynomials in the context of polynomial optimization. More concretely, a symmetric matrix $\mathbf{G} \in S_t$ is *diagonally dominant* if $\mathbf{G}_{ii} \geq \sum_{j \neq i} |\mathbf{G}_{ij}|$ for $i \in [t]$, and is *scaled diagonally dominant* if there exists a positive definite $t \times t$ diagonal matrix $\mathbf{D}$ such that $\mathbf{DGD}$ is diagonally dominant. We say that a polynomial $f \in \mathbb{R}[\mathbf{x}]$ is a *scaled diagonally dominant sum of squares* (SDSOS) polynomial if it admits a Gram matrix representation as in (7.9) (with $b = 0$) with a scaled diagonally dominant Gram matrix $\mathbf{G}$. We denote the set of SDSOS polynomials by SDSOS.

By replacing the nonnegativity condition in **P** with the SDSOS condition, one obtains the SDSOS relaxation of **P**:

$$(\text{SDSOS}): \quad f_{\text{sdsos}} := \sup \{b : f - b \in \text{SDSOS}\}.$$

It turns out that the first value of the TSSOS hierarchy is already better than or equal to the bound provided by the SDSOS relaxation.

Theorem 7.8 *With the above notation, one has $f_{\text{ts}}^1 \geq f_{\text{sdsos}}$.*

PROOF Let $\mathscr{A} = \text{supp}(f)$ and $\mathscr{B}$ be a monomial basis. Assume $f \in$ SDSOS, i.e., f admits a scaled diagonally dominant Gram matrix $\mathbf{G} \in S_{|\mathscr{B}|}^+$ indexed by $\mathscr{B}$. We then construct a Gram matrix $\tilde{\mathbf{G}}$ for f by

$$\tilde{\mathbf{G}}_{\beta\gamma} = \begin{cases} \mathbf{G}_{\beta\gamma}, & \text{if } \beta + \gamma \in \mathscr{A} \cup (2\mathscr{B}), \\ 0, & \text{otherwise.} \end{cases}$$

It is easy to see that we still have $f = (\mathbf{x}^{\mathscr{B}})^\intercal \tilde{\mathbf{G}} \mathbf{x}^{\mathscr{B}}$. Note that we only replace off-diagonal entries by zeros in $\mathbf{G}$ to obtain $\tilde{\mathbf{G}}$ and replacing off-diagonal entries by zeros does not affect the scaled diagonal dominance of a matrix. Hence $\tilde{\mathbf{G}}$ is also a scaled diagonally dominant matrix. Moreover, we have $\tilde{\mathbf{G}} \in S_{|\mathscr{B}|}^+ \cap S(G^{(1)})$ by construction. It follows that (SDSOS) is a relaxation of (7.9). Hence $f_{\text{ts}}^1 \geq f_{\text{sdsos}}$ by Proposition 7.7.   □

7.3   The TSSOS hierarchy for constrained POPs

In this section, we describe an iterative procedure to exploit TS for the primal-dual moment-SOS hierarchy of constrained POPs:

$$\mathbf{P}: \quad f_{\min} := \inf \{f(\mathbf{x}) : \mathbf{x} \in \mathbf{X}\}, \tag{7.10}$$

with

$$\mathbf{X} = \{\mathbf{x} \in \mathbb{R}^n \mid g_1(\mathbf{x}) \geq 0, \dots, g_m(\mathbf{x}) \geq 0\}. \tag{7.11}$$

Let $\mathscr{A}$ denote the union of supports involved in $\mathbf{P}$, i.e.,

$$\mathscr{A} = \mathrm{supp}(f) \cup \bigcup_{j=1}^{m} \mathrm{supp}(g_j). \tag{7.12}$$

Let $r_{\min} := \max \{\lceil \deg(f)/2 \rceil, d_1, \ldots, d_m\}$ with $d_j := \lceil \deg(g_j)/2 \rceil$ for $j \in [m]$. Fix a relaxation order $r \geq r_{\min}$. Let $g_0 = 1$, $d_0 = 0$ and $\mathscr{B}_{r,j} = \mathbb{N}^n_{r-d_j}$ be the standard monomial basis for $j = 0, 1, \ldots, m$. We define a graph G^{tsp} with nodes $\mathscr{B}_{r,0}$ and edges

$$E(G^{\mathrm{tsp}}) = \{\{\beta, \gamma\} \mid \beta \neq \gamma \in \mathscr{B}_{r,0}, \ \beta + \gamma \in \mathscr{A} \cup (2\mathscr{B}_{r,0})\}, \tag{7.13}$$

which is called the tsp graph associated with $\mathbf{P}$ or essentially $\mathscr{A}$.

Now let us initialize with $G^{(0)}_{r,0} := G^{\mathrm{tsp}}$ and $G^{(0)}_{r,j}$ being an empty graph for $j \in [m]$. Then for each $j \in \{0\} \cup [m]$, we define a sequence of graphs $(G^{(s)}_{r,j})_{s \geq 1}$ by iteratively performing two successive operations:

(1) **support extension**. Let $F^{(s)}_{r,j}$ be the graph with nodes $\mathscr{B}_{r,j}$ and edges

$$E(F^{(s)}_{r,j}) = \Big\{\{\beta, \gamma\} \mid \beta \neq \gamma \in \mathscr{B}_{r,j},$$
$$(\mathrm{supp}(g_j) + \beta + \gamma) \cap \bigcup_{i=0}^{m} (\mathrm{supp}(g_i) + \mathrm{supp}(G^{(s-1)}_{r,i})) \neq \varnothing\Big\}. \tag{7.14}$$

(2) **chordal extension**. Let

$$G^{(s)}_{r,j} := (F^{(s)}_{r,j})', \quad j \in \{0\} \cup [m]. \tag{7.15}$$

Recall that the dense moment relaxation of order r for $\mathbf{P}$ is given by

$$\mathbf{P}^r: \quad \begin{aligned} f^r_{\mathrm{mom}} := \ &\inf_{\mathbf{y}} \ L_{\mathbf{y}}(f) \\ &\text{s.t.} \ \ \mathbf{M}_{r-d_j}(g_j\,\mathbf{y}) \succeq 0, \quad j \in \{0\} \cup [m] \\ &\quad\quad y_0 = 1. \end{aligned} \tag{7.15}$$

Let $t_j := |\mathscr{B}_{r,j}| = \binom{n+r-d_j}{r-d_j}$. Therefore by replacing $\mathbf{M}_{r-d_j}(g_j\mathbf{y}) \succeq 0$ with the weaker condition $\mathbf{B}_{G^{(s)}_{r,j}} \circ \mathbf{M}_{r-d_j}(g_j\mathbf{y}) \in \Pi_{G^{(s)}_{r,j}}(S^+_{t_j})$ for $j \in \{0\} \cup [m]$ in (7.16), we obtain the following sparse moment relaxation of $\mathbf{P}^r$ and $\mathbf{P}$ for each $s \geq 1$:

$$\mathbf{P}^{r,s}_{\mathrm{ts}}: \quad \begin{aligned} f^{r,s}_{\mathrm{ts}} := \ &\inf_{\mathbf{y}} \ L_{\mathbf{y}}(f) \\ &\text{s.t.} \ \ \mathbf{B}_{G^{(s)}_{r,0}} \circ \mathbf{M}_r(\mathbf{y}) \in \Pi_{G^{(s)}_{r,0}}(S^+_{t_0}) \\ &\quad\quad \mathbf{B}_{G^{(s)}_{r,j}} \circ \mathbf{M}_{r-d_j}(g_j\mathbf{y}) \in \Pi_{G^{(s)}_{r,j}}(S^+_{t_j}), \quad j \in [m] \\ &\quad\quad y_0 = 1. \end{aligned} \tag{7.17}$$

As in the unconstrained case, we call s the sparse order. By construction, one has $G_{r,j}^{(s)} \subseteq G_{r,j}^{(s+1)}$ for all j,s. Therefore, for each $j \in \{0\} \cup [m]$, the sequence of graphs $(G_{r,j}^{(s)})_{s \geq 1}$ stabilizes after a finite number of steps. We denote the stabilized graph by $G_{r,j}^{(\bullet)}$ for all j and the corresponding moment relaxation by $\mathbf{P}_{ts}^{r,\bullet}$ with optimum $f_{ts}^{r,\bullet}$.

For each $s \geq 1$, the dual SDP of $\mathbf{P}_{ts}^{r,s}$ reads as

$$
\begin{cases}
\sup & b \\
\mathbf{G}_{j,b} & \\
\text{s.t.} & \sum_{j=0}^{m} \langle \mathbf{C}_{\alpha}^{j}, \mathbf{G}_j \rangle = f_{\alpha} - b\mathbf{1}_{\alpha=0}, \quad \forall \alpha \in \bigcup_{j=0}^{m}(\text{supp}(g_j) + \text{supp}(G_{r,j}^{(s)})) \\
& \mathbf{G}_j \in \mathbf{S}_{t_j}^{+} \cap \mathbf{S}(G_{r,j}^{(s)}), \quad j \in \{0\} \cup [m]
\end{cases}
$$

$$(7.18)$$

where $\mathbf{C}_{\alpha}^{j}$ is defined after (2.8). The primal-dual SDP relaxations (7.17)–(7.18) are called the TSSOS hierarchy associated with $\mathbf{P}$, which is indexed by two parameters: the relaxation order r and the sparse order s.

Theorem 7.9 *With the above notation, the following hold:*

(i) *Assume that* $\mathbf{X}$ *has a nonempty interior. Then there is no duality gap between* $\mathbf{P}_{ts}^{r,s}$ *and its dual (7.18) for any* $r \geq r_{\min}$ *and* $s \geq 1$.

(ii) *Fixing a relaxation order* $r \geq r_{\min}$, *the sequence* $(f_{ts}^{r,s})_{s \geq 1}$ *is monotonically nondecreasing and* $f_{ts}^{r,s} \leq f_{\text{mom}}^{r}$ *for all* $s \geq 1$.

(iii) *When the maximal chordal extension is used for the chordal extension operation, the sequence* $(f_{ts}^{r,s})_{s \geq 1}$ *converges to* f_{mom}^{r} *in finitely many steps.*

(iv) *Fixing a sparse order* $s \geq 1$, *the sequence* $(f_{ts}^{r,s})_{r \geq r_{\min}}$ *is monotonically nondecreasing.*

PROOF (i). This easily follows from the fact that $\mathbf{P}_{ts}^{r,s}$ satisfies Slater's condition by Theorem 4.2 of [Las01] and Theorem 1.5.

(ii). For all j,s, the inclusion $G_{r,j}^{(s)} \subseteq G_{r,j}^{(s+1)}$ implies that each maximal clique of $G_{r,j}^{(s)}$ is a subset of some maximal clique of $G_{r,j}^{(s+1)}$. Hence by Theorem 1.5, $\mathbf{P}_{ts}^{r,s}$ is a relaxation of $\mathbf{P}_{ts}^{r,s+1}$ (and also a relaxation of $\mathbf{P}^r$) from which we have that $(f_{ts}^{r,s})_{s \geq 1}$ is monotonically nondecreasing and $f_{ts}^{r,s} \leq f_{\text{mom}}^{r}$ for all $s \geq 1$.

(iii). Let $\mathbf{y} = (y_{\alpha})$ be an arbitrary feasible solution of $\mathbf{P}_{ts}^{r,\bullet}$. We note that

$\{y_\alpha \mid \alpha \in \bigcup_{j \in \{0\} \cup [m]} (\operatorname{supp}(g_j) + \operatorname{supp}(G_{r,j}^{(\bullet)}))\}$ is the set of decision variables involved in $\mathbf{P}_{ts}^{r,\bullet}$, and $\{y_\alpha \mid \alpha \in \mathbb{N}_{2r}^n\}$ is the set of decision variables involved in $\mathbf{P}^r$ (7.16). We then define a vector $\bar{y} = (\bar{y}_\alpha)_{\alpha \in \mathbb{N}_{2r}^n}$ as follows:

$$\bar{y}_\alpha = \begin{cases} y_\alpha, & \text{if } \alpha \in \bigcup_{j \in \{0\} \cup [m]} (\operatorname{supp}(g_j) + \operatorname{supp}(G_{r,j}^{(\bullet)})), \\ 0, & \text{otherwise.} \end{cases}$$

By construction and since $G_{r,j}^{(\bullet)}$ stabilizes under support extension for all j, we immediately have $\mathbf{M}_{r-d_j}(g_j\bar{y}) = \mathbf{B}_{G_{r,j}^{(\bullet)}} \circ \mathbf{M}_{r-d_j}(g_j y)$. As we use the maximal chordal extension for the chordal extension operation, the matrix $\mathbf{B}_{G_{r,j}^{(\bullet)}} \circ \mathbf{M}_{r-d_j}(g_j y)$ is block-diagonal up to permutation. So from $\mathbf{B}_{G_{r,j}^{(\bullet)}} \circ \mathbf{M}_{r-d_j}(g_j y) \in \Pi_{G_{r,j}^{(\bullet)}}(\mathbf{S}_+^{t_j})$ it follows $\mathbf{M}_{r-d_j}(g_j\bar{y}) \succeq 0$ for $j \in \{0\} \cup [m]$. Therefore $\bar{y}$ is a feasible solution of $\mathbf{P}^r$ and so $L_y(f) = L_{\bar{y}}(f) \geq f_{mom}^r$. Hence $f_{ts}^{r,\bullet} \geq f_{mom}^r$ as y is an arbitrary feasible solution of $\mathbf{P}_{ts}^{r,\bullet}$. By (ii), we already have $f_{ts}^{r,\bullet} \leq f_{mom}^r$. Therefore, $f_{ts}^{r,\bullet} = f_{mom}^r$ as desired.

(iv). The conclusion follows if we can show that $G_{r,j}^{(s)} \subseteq G_{r+1,j}^{(s)}$ for all j, r since by Theorem 1.5 this implies that $\mathbf{P}_{ts}^{r,s}$ is a relaxation of $\mathbf{P}_{ts}^{r+1,s}$. Let us prove $G_{r,j}^{(s)} \subseteq G_{r+1,j}^{(s)}$ by induction on s. For $s = 1$, from (7.13), we have $G_{r,0}^{(0)} \subseteq G_{r+1,0}^{(0)}$, which implies $G_{r,j}^{(1)} \subseteq G_{r+1,j}^{(1)}$ for $j \in \{0\} \cup [m]$. Now assume that $G_{r,j}^{(s)} \subseteq G_{r+1,j}^{(s)}, j \in \{0\} \cup [m]$ hold for a given $s \geq 1$. Then from (7.14) and by the induction hypothesis, we have $G_{r,j}^{(s+1)} \subseteq G_{r+1,j}^{(s+1)}$ for $j \in \{0\} \cup [m]$, which completes the induction and also completes the proof. $\square$

By Theorem 7.9, we have the following two-level hierarchy of lower bounds for the optimum f_{min} of $\mathbf{P}$:

$$\begin{array}{ccccccc}
f_{ts}^{r_{min},1} & \leq & f_{ts}^{r_{min},2} & \leq & \cdots & \leq & f_{mom}^{r_{min}} \\
\wedge | & & \wedge | & & & & \wedge | \\
f_{ts}^{r_{min}+1,1} & \leq & f_{ts}^{r_{min}+1,2} & \leq & \cdots & \leq & f_{mom}^{r_{min}+1} \\
\wedge | & & \wedge | & & & & \wedge | \\
\vdots & & \vdots & & \vdots & & \vdots \\
\wedge | & & \wedge | & & & & \wedge | \\
f_{ts}^{r,1} & \leq & f_{ts}^{r,2} & \leq & \cdots & \leq & f_{mom}^{r} \\
\wedge | & & \wedge | & & & & \wedge | \\
\vdots & & \vdots & & \vdots & &
\end{array} \tag{7.19}$$

The TSSOS hierarchy entails a trade-off between the computational cost and the quality of the obtained lower bound via the two tunable parameters r and s. Besides, one has the freedom to choose a specific chordal extension in (7.15) (e.g., maximal chordal extensions, approximately smallest chordal extensions and so on). This choice could affect the resulting sizes of PSD blocks and the quality of the related lower bound. Intuitively, chordal extensions with smaller clique numbers would lead to PSD blocks of smaller sizes and lower bounds of (possibly) lower quality while chordal extensions with larger clique numbers would lead to PSD blocks with larger sizes and lower bounds of (possibly) higher quality.

Remark 7.10 *If* $\mathbf{P}$ *is a QCQP, then* $\mathbf{P}_{ts}^{1,1}$ *and* $\mathbf{P}^1$ *yield the same lower bound, i.e.,* $f_{ts}^{1,1} = f_{mom}^1$. *Indeed, for a QCQP, the moment relaxation* $\mathbf{P}^1$ *reads as*

$$\begin{cases} \inf_{\mathbf{y}} & L_{\mathbf{y}}(f) \\ \text{s.t.} & \mathbf{M}_1(\mathbf{y}) \succeq 0 \\ & L_{\mathbf{y}}(g_j) \geq 0, \quad j \in [m] \\ & y_0 = 1. \end{cases}$$

Note that the objective function and the affine constraints of $\mathbf{P}^1$ *involve only the decision variables* $\{y_0\} \cup \{y_\alpha\}_{\alpha \in \mathscr{A}}$ *with* $\mathscr{A} = \operatorname{supp}(f) \cup \bigcup_{j=1}^m \operatorname{supp}(g_j)$. *Hence there is no discrepancy of optima in replacing* $\mathbf{P}^1$ *with* $\mathbf{P}_{ts}^{1,1}$ *by construction.*

7.4   Obtaining a possibly smaller monomial basis

The size of SDPs arising form the TSSOS hierarchy heavily depends on the chosen monomial basis $\mathscr{B}$ or $\mathscr{B}_{r,0}$. As we already saw in Section 7.1, for unconstrained POPs the Newton polytope method usually produces a monomial basis smaller than the standard monomial basis. However, this method does not apply to constrained POPs. Here as an optional pre-treatment, we present an iterative procedure which not only allows us to obtain a monomial basis smaller than the one given by the Newton polytope method for unconstrained POPs in many cases, but can also be applied to constrained POPs.

We start with the unconstrained case. Let $f \in \mathbb{R}[\mathbf{x}]$ with $\mathscr{A} = \operatorname{supp}(f)$ and $\mathscr{B}$ be the monomial basis given by the Newton polytope method. Initializing with $\mathscr{B}_0 := \varnothing$, we iteratively define a sequence of monomial sets $(\mathscr{B}_p)_{p \geq 1}$ by

$$\mathscr{B}_p := \{\beta \in \mathscr{B} \mid \exists \gamma \in \mathscr{B} \text{ s.t. } \beta + \gamma \in \mathscr{A} \cup (2\,\mathscr{B}_{p-1})\}. \tag{7.20}$$

Consequently, we obtain an ascending chain of monomial sets:

$$\mathscr{B}_1 \subseteq \mathscr{B}_2 \subseteq \mathscr{B}_3 \subseteq \cdots \subseteq \mathscr{B}.$$

This procedure is formulated in Algorithm 4. Each $\mathscr{B}_p$ in the chain can serve as a candidate monomial basis. In practice, if indexing the unknown Gram matrix by $\mathscr{B}_p$ leads to an infeasible SDP, then we turn to $\mathscr{B}_{p+1}$ until a feasible SDP is retrieved.

Algorithm 4 GenerateBasis

Require: A support set $\mathscr{A}$ and an initial monomial basis $\mathscr{B}$
Ensure: An ascending chain of potential monomial bases $(\mathscr{B}_p)_{p\geq 1}$
1: $\mathscr{B}_0 \leftarrow \varnothing$
2: $p \leftarrow 0$
3: **while** $p = 0$ or $\mathscr{B}_p \neq \mathscr{B}_{p-1}$ **do**
4:      $p \leftarrow p + 1$
5:      $\mathscr{B}_p \leftarrow \varnothing$
6:      **for** each pair β, γ in $\mathscr{B}$ **do**
7:          **if** $\beta + \gamma \in \mathscr{A} \cup (2\mathscr{B}_{p-1})$ **then**
8:              $\mathscr{B}_p \leftarrow \mathscr{B}_p \cup \{\beta, \gamma\}$
9:          **end if**
10:     **end for**
11: **end while**
12: **return** $(\mathscr{B}_p)_{p\geq 1}$

Proposition 7.11 *Let $f \in \mathbb{R}[\mathbf{x}]$ and $\mathscr{B}_* = \cup_{p\geq 1} \mathscr{B}_p$ with $\mathscr{B}_p$ being defined by (7.20). If $f \in$ SDSOS, then f is an SDSOS polynomial in the monomial basis $\mathscr{B}_*$.*

PROOF Let $\mathscr{B}$ be the monomial basis returned by the Newton polytope method. If $f \in$ SDSOS, then there exists a scaled diagonally dominant Gram matrix $\mathbf{G} \in \mathbb{S}_+^{|\mathscr{B}|}$ indexed by $\mathscr{B}$ such that $f = (\mathbf{x}^{\mathscr{B}})^{\mathsf{T}}\mathbf{G}\mathbf{x}^{\mathscr{B}}$. We then construct a Gram matrix $\tilde{\mathbf{G}} \in \mathbb{S}_+^{|\mathscr{B}_*|}$ indexed by $\mathscr{B}_*$ for f as follows:

$$\tilde{\mathbf{G}}_{\beta\gamma} = \begin{cases} \mathbf{G}_{\beta\gamma}, & \text{if } \beta + \gamma \in \mathscr{A} \cup (2\mathscr{B}_*), \\ 0, & \text{otherwise.} \end{cases}$$

One can easily check that we still have $f = (\mathbf{x}^{\mathscr{B}_*})^{\mathsf{T}}\tilde{\mathbf{G}}\mathbf{x}^{\mathscr{B}_*}$. Let $\hat{\mathbf{G}}$ be the principal submatrix of $\mathbf{G}$ by deleting the rows and columns whose indices are not in $\mathscr{B}_*$, which is also a scaled diagonally dominant matrix. By construction, $\tilde{\mathbf{G}}$ is obtained from $\hat{\mathbf{G}}$ by replacing certain off-diagonal entries by zeros. Since replacing off-diagonal entries by zeros does not affect the scaled diagonal dominance of a matrix, $\tilde{\mathbf{G}}$ is also a scaled diagonally dominant matrix. It follows that f is an SDSOS polynomial in the monomial basis $\mathscr{B}_*$. $\qquad\square$

Remark 7.12 *By Proposition 7.11, if we use the monomial basis $\mathcal{B}_*$ for (7.7)–(7.9), we still have the hierarchy of optima:*

$$f_{\text{sdsos}} \leq f_{\text{ts}}^1 \leq f_{\text{ts}}^2 \leq \cdots \leq f_{\text{mom}} \leq f_{\text{min}}.$$

The algorithm `GenerateBasis` may provide a smaller monomial basis than the one given by the Newton polytope method as the following example illustrates.

Example 7.13 *Consider the polynomial $f = 1 + x + x^8$. The monomial basis given by the Newton polytope method is $\mathcal{B} = \{1, x, x^2, x^3, x^4\}$. By the algorithm `GenerateBasis`, we obtain $\mathcal{B}_1 = \{1, x, x^4\}$ and $\mathcal{B}_2 = \{1, x, x^2, x^4\}$. It turns out that f admits no SOS decomposition with $\mathcal{B}_1$ while $\mathcal{B}_2$ can serve as a monomial basis to represent f as an SOS.*

For the constrained case we follow the notation of Section 7.3. Fix a relaxation order r and a sparse order s of the TSSOS hierarchy. Initializing with $\mathcal{B}_{r,0} = \mathbb{N}_r^n$, we iteratively perform the following two steps:

For Step 1, let the maximal cliques of $G_{r,j}^{(s)}$ be $C_{j,1}, C_{j,2}, \ldots, C_{j,t_j}$ for $j \in \{0\} \cup [m]$. Let

$$\mathscr{F} = \text{supp}(f) \cup \bigcup_{j=1}^{m} \left(\text{supp}(g_j) + \bigcup_{i=1}^{t_j} (C_{j,i} + C_{j,i}) \right). \qquad (7.21)$$

Then call the algorithm `GenerateBasis` with $\mathscr{A} = \mathscr{F}$ and $\mathscr{B} = \mathscr{B}_{r,0}$ to generate a new monomial basis $\mathscr{B}'_{r,0}$.

For Step 2, with the new monomial basis $\mathscr{B}'_{r,0}$, recompute the graph $G_{r,j}^{(s)}$ for $j \in \{0\} \cup [m]$ as in Section 7.3 and then go back to Step 1.

Continue the iterative procedure until $\mathscr{B}'_{r,0} = \mathscr{B}_{r,0}$, which is the desired monomial basis.

7.5   Sign symmetries and a sparse representation theorem for positive polynomials

The exploitation of TS developed in the previous sections is closely related to *sign symmetries*. Intuitively, a polynomial is said to have sign symmetries if it is invariant when we change signs of some variables. For instance, the polynomial $f(x_1, x_2) = x_1^2 + x_2^2 + x_1 x_2$ has the sign symmetry associated to $(x_1, x_2) \mapsto (-x_1, -x_2)$ as $f(-x_1, -x_2) = f(x_1, x_2)$. To be more precise, we give the following definition of sign symmetries in terms of support sets.

Definition 7.14 (sign symmetry) *Given a finite set $\mathscr{A} \subseteq \mathbb{N}^n$, the sign symmetries of $\mathscr{A}$ are defined by all vectors $\mathbf{s} \in \mathbb{Z}_2^n := \{0,1\}^n$ such that $\mathbf{s}^\top \alpha \equiv 0$ (mod 2) for all $\alpha \in \mathscr{A}$.*

Assume that the maximal chordal extension is chosen for the chordal extension operation in Section 7.3. As mentioned earlier, for any j the sequence of graphs $(G_{r,j}^{(s)})_{s \geq 1}$ ends up with $G_{r,j}^{(\bullet)}$ in finitely many steps. Note that the graph $G_{r,j}^{(\bullet)}$ induces a partition of the monomial basis $\mathbb{N}_{r-d_j}^n$: two monomials $\beta, \gamma \in \mathbb{N}_{r-d_j}^n$ belong to the same block if and only if they belong to the same connected component of $G_{r,j}^{(\bullet)}$. The following theorem provides an interpretation of this partition in terms of sign symmetries.

Theorem 7.15 *Notations are as in the previous sections. Fix the relaxation order $r \geq r_{\min}$. Assume that the maximal chordal extension is chosen for the chordal extension operation and the sign symmetries of $\mathscr{A}$ are given by the columns of a binary matrix denoted by $\mathbf{R}$. Then for each $j \in \{0\} \cup [m]$, β, γ belong to the same block in the partition of $\mathbb{N}_{r-d_j}^n$ induced by $G_{r,j}^{(\bullet)}$ if and only if $\mathbf{R}^\top(\beta + \gamma) \equiv 0$ (mod 2). In other words, for a fixed relaxation order the block structures arising from the TSSOS hierarchy converge to the block structure determined by the sign symmetries of the POP assuming that the maximal chordal extension is used for the chordal extension operation.*

Theorem 7.15 is applied for the standard monomial basis $\mathbb{N}_{r-d_j}^n$. If a smaller monomial basis is chosen, then we only have the "only if" part of the conclusion in Theorem 7.15.

Example 7.16 *Let $f = 1 + x_1^2 x_2^4 + x_1^4 x_2^2 + x_1^4 x_2^4 - x_1 x_2^2 - 3 x_1^2 x_2^2$ and $\mathscr{A} = \mathrm{supp}(f)$. The monomial basis returned by the Newton polytope method is $\mathscr{B} = \{1, x_1 x_2, x_1 x_2^2, x_1^2 x_2, x_1^2 x_2^2\}$. The sign symmetries of $\mathscr{A}$ consist of two elements: $(0,0)$ and $(0,1)$. According to the sign symmetries, the basis $\mathscr{B}$ is partitioned into $\{1, x_1 x_2^2, x_1^2 x_2^2\}$ and $\{x_1 x_2, x_1^2 x_2\}$. On the other hand, the partition of $\mathscr{B}$ induced by $G^{(\bullet)}$ is $\{1, x_1 x_2^2, x_1^2 x_2^2\}$, $\{x_1 x_2\}$ and $\{x_1^2 x_2\}$, which is a refinement of the partition determined by the sign symmetries.*

As a corollary of Theorem 7.15, we can prove a sparse representation theorem for positive polynomials on compact basic semialgebraic sets.

Theorem 7.17 *Let* $\mathbf{X}$ *be defined as in* (7.11). *Assume that the quadratic module* $\mathcal{M}(\mathfrak{g})$ *is Archimedean and that the polynomial* f *is positive on* $\mathbf{X}$. *Let* $\mathscr{A} = \mathrm{supp}(f) \cup \bigcup_{j=1}^m \mathrm{supp}(g_j)$ *and let the sign symmetries of* $\mathscr{A}$ *be given by the columns of the binary matrix* R. *Then* f *can be decomposed as*

$$f = \sigma_0 + \sum_{j=1}^m \sigma_j g_j,$$

for some SOS polynomials $\sigma_0, \sigma_1, \ldots, \sigma_m$ *satisfying* $R^\mathsf{T} \alpha \equiv 0 \pmod 2$ *for any* $\alpha \in \mathrm{supp}(\sigma_j), j = 0, \ldots, m$.

PROOF By Putinar's Positivstellensatz (Theorem 2.3), there exist SOS polynomials $\sigma_0', \sigma_1', \ldots, \sigma_m'$ such that

$$f = \sigma_0' + \sum_{j=1}^m \sigma_j' g_j. \tag{7.22}$$

Let $d_j = \lceil \deg(g_j)/2 \rceil, j = 0, 1, \ldots, m$ and

$$r = \max \{ \lceil \deg(\sigma_j' g_j)/2 \rceil : j = 0, 1, \ldots, m \}$$

with $g_0 = 1$. Let $\mathbf{G}_j$ be a Gram matrix associated with σ_j' and indexed by the monomial basis $\mathbb{N}_{r-d_j}^n, j \in \{0\} \cup [m]$. We define $\sigma_j = (\mathbf{x}^{\mathbb{N}_{r-d_j}^n})^\mathsf{T}(\mathbf{B}_{G_{r,j}^{(\bullet)}} \circ \mathbf{G}_j)\mathbf{x}^{\mathbb{N}_{r-d_j}^n}$ for $j = 0, 1, \ldots, m$, where $G_{r,j}^{(\bullet)}$ is defined in Section 7.3. For any $j \in \{0\} \cup [m]$, since $\mathbf{B}_{G_{r,j}^{(\bullet)}} \circ \mathbf{G}_j$ is block-diagonal (up to permutation) and $\mathbf{G}_j$ is positive semidefinite, we see that σ_j is an SOS polynomial.

Suppose $\alpha \in \mathrm{supp}(\sigma_j)$ with $j \in \{0\} \cup [m]$. Then we can write $\alpha = \alpha' + \beta + \gamma$ for some $\alpha' \in \mathrm{supp}(g_j)$ and some β, γ belonging to the same connected component of $G_{r,j}^{(\bullet)}$. By Theorem 7.15, we have $\mathbf{R}^\mathsf{T}(\beta + \gamma) \equiv 0 \pmod 2$ and therefore, $\mathbf{R}^\mathsf{T}\alpha \equiv 0 \pmod 2$. Moreover, for any $\alpha' \in \mathrm{supp}(g_j)$ and β, γ not belonging to the same connected component of $G_{r,j}^{(\bullet)}$, we have $\mathbf{R}^\mathsf{T}(\beta + \gamma) \not\equiv 0 \pmod 2$ by Theorem 7.15 and so $\mathbf{R}^\mathsf{T}(\alpha' + \beta + \gamma) \not\equiv 0 \pmod 2$. From these facts we deduce that substituting σ_j' with σ_j in (7.22) is just removing the terms whose exponents α do not satisfy $\mathbf{R}^\mathsf{T}\alpha \equiv 0 \pmod 2$ from the right-hand side of (7.22). Doing so, one does not change the match of coefficients on both sides of the equality. Thus we have

$$f = \sigma_0 + \sum_{j=1}^m \sigma_j g_j,$$

with the desired property. □

7.6   Numerical experiments

We present some numerical results of the proposed TSSOS hierarchy in this section. The related algorithms were implemented in the Julia package TSSOS.[1] For the numerical experiments, we use Mosek [ART03] as an SDP solver. All examples were computed on an Intel Core i5-8265U@1.60GHz CPU with 8GB RAM memory. The timing (in seconds) includes the time for pre-processing (to get the block structure), the time for modeling SDP and the time for solving SDP.

7.6.1   Unconstrained polynomial optimization

We first consider Lyapunov functions emerging from some networked systems. The following polynomial is from Example 2 in [HP11]:

$$f = \sum_{i=1}^{n} a_i(x_i^2 + x_i^4) - \sum_{i=1}^{n}\sum_{k=1}^{n} b_{ik}x_i^2 x_k^2,$$

where a_i are randomly chosen from $[1, 2]$ and b_{ik} are randomly chosen from $[\frac{0.5}{n}, \frac{1.5}{n}]$. Here, n is the number of nodes in the network. The task is to determine whether f is globally nonnegative.

The sizes of SDPs corresponding to the TSSOS (with sparse order $s = 1$) and dense relaxations are listed in Table 7.1. In the column "#PSD blocks", $i \times j$ means j PSD blocks of size i. The column "#equality constraints" indicates the number of equality constraints involved in SDPs.

Table 7.1: The sizes of SDPs for the sparse and dense relaxations.

	#PSD blocks	#equality constraints
TSSOS	$3 \times \frac{n(n-1)}{2}, 1 \times n, (n+1) \times 1$	$\frac{3n(n-1)}{2} + 2n + 1$
Dense	$\binom{n+2}{2} \times 1$	$\binom{n+4}{4}$

We solve the TSSOS relaxation with $s = 1$ for $n \in \{10, 20, \ldots, 80\}$. The results are displayed in Table 7.2 in which "mb" stands for the maximal size of PSD blocks.

Table 7.2: Results for the first networked system.

n	10	20	30	40	50	60	70	80
mb	11	31	31	41	51	61	71	81
time	0.006	0.03	0.10	0.34	0.92	1.9	4.7	12

---

[1]https://github.com/wangjie212/TSSOS.

For this example, the size of the system that can be handled in [HP11] is up to $n = 50$ nodes while TSSOS can easily handle the system with up to $n = 80$ nodes.

The following polynomial is from Example 3 in [HP11]:

$$V = \sum_{i=1}^{n} a_i \left(\frac{1}{2} x_i^2 - \frac{1}{4} x_i^4 \right) + \frac{1}{2} \sum_{i=1}^{n} \sum_{k=1}^{n} b_{ik} \frac{1}{4} (x_i - x_k)^4, \qquad (7.23)$$

where a_i are randomly chosen from $[0.5, 1.5]$ and b_{ik} are randomly chosen from $[\frac{0.5}{n}, \frac{1.5}{n}]$. The task is to analyze the domain on which the Hamiltonian function V for a network of Duffing oscillators is positive definite. We use the following condition to establish an inner approximation of the domain on which V is positive definite:

$$f = V - \sum_{i=1}^{n} \lambda_i x_i^2 (g - x_i^2) \geq 0, \qquad (7.24)$$

where $\lambda_i > 0$ are scalar decision variables and g is a fixed positive scalar. Clearly, the condition (7.24) ensures that V is positive definite when $x_i^2 < g$.

We illustrate the tsp graph $G^{(0)}$ of this example in Figure 7.4, which has 1 maximal clique of size $n + 1$ (involving the nodes $1, x_1^2, \ldots, x_n^2$), $\frac{n(n-1)}{2}$ maximal cliques of size 3 (involving the nodes $x_i^2, x_j^2, x_i x_j$ for each pair $\{i, j\}, i \neq j$) and n maximal cliques of size 1 (involving the node x_i for each i). Since $G^{(0)}$ is already a chordal graph, we have $G^{(1)} = G^{(0)}$.

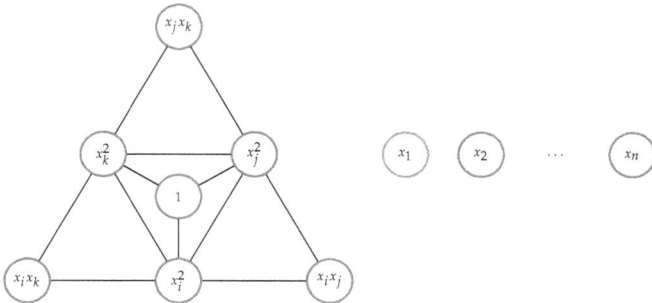

Figure 7.4: The tsp graph $G^{(0)}$ for the second networked system. What is displayed is merely a subgraph of $G^{(0)}$. The whole graph $G^{(0)}$ can be obtained by putting all such subgraphs together.

Here we solve the TSSOS relaxation with $s = 1$ for $n \in \{10, 20, \ldots, 50\}$. This example was also examined in [MAT14] to demonstrate the advantage of SDSOS programming compared to dense SOS programming. The

SDSOS approach was implemented in SPOT [Meg10] with Mosek as a second-order cone programming solver. We display the results for both TSSOS and SDSOS in Table 7.3 in which the row "#SDP vars" indicates the number of decision variables used in SDPs.

Table 7.3: Results for the second networked system.

n		10	20	30	40	50
#PSD blocks	TSSOS	3×45, 1×10, 11×1	3×190, 1×20, 21×1	3×435, 1×30, 31×1	3×780, 1×40, 41×1	3×1225, 1×50, 51×1
	SDSOS	2×2145	2×26565	2×122760	2×370230	2×878475
#SDP vars	TSSOS	346	1391	3136	5581	8726
	SDSOS	6435	79695	368280	1110690	2635425
time	TSSOS	0.01	0.06	0.17	0.50	0.89
	SDSOS	0.47	1.14	5.47	20	70

From the table we see that the TSSOS approach uses much less decision variables than the SDSOS approach, and hence is more efficient. On the other hand, the TSSOS approach computes a positive definite form V after selecting a value for g up to 2 (which is the same as the maximal value obtained by the dense SOS approach) while the method in [HP11] can select g up to 1.8 and the SDSOS approach only works out for a maximal value of g up to around 1.5.

7.6.2   Constrained polynomial optimization

Now we present the numerical results for constrained polynomial optimization problems. As a first example, we minimize randomly generated[2] sparse polynomials H_2, H_4, H_6 over the unit ball

$$\mathbf{X} = \left\{ (x_1, \ldots, x_n) \in \mathbb{R}^n \,\middle|\, g_1 = 1 - (x_1^2 + \cdots + x_n^2) \geq 0 \right\}.$$

We solve these instances with TSSOS as well as GloptiPoly. The related results are shown in Table 7.4. Note that approximately smallest chordal extensions are used for TS and only the results of the first three steps of the TSSOS hierarchy (for a fixed relaxation order) are displayed. From the table it can be seen that for each instance TSSOS is significantly faster than GloptiPoly[3] without compromising accuracy.

Next we present the numerical results (Table 7.5) for minimizing the

---

[2]These polynomials are available at https://wangjie212.github.io/jiewang/code.html.
[3]GloptiPoly also uses Mosek as an SDP solver.

Table 7.4: Results for minimizing randomly generated polynomials over the unit ball; d denotes the polynomial degree and t denotes the number of terms.

(n,d,t)		r	TSSOS				GloptiPoly		
			s	mb	opt	time	mb	opt	time
H_2	$(7,8,12)$	4	1	36		0.36	330	0.1373	34
			2	36		0.52			
			3	38	0.1373	1.6			
		5	1	36		1.9	792	-	-
			2	45		3.9			
			3	59		34			
H_4	$(9,6,15)$	3	1	10		0.15	220	0.8704	16
			2	10		0.22			
			3	10	0.8704	0.25			
		4	1	55		1.3	715	-	-
			2	55		2.0			
			3	56		2.8			
H_6	$(11,6,20)$	3	1	12		0.28	364	0.1171	115
			2	15		0.36			
			3	16	0.1171	0.60			
		4	1	78		4.4	1365	-	-
			2	78		4.7			
			3	78		7.5			

generalized Rosenbrock function over the unit ball:

$$f_{\text{gR}}(\mathbf{x}) = 1 + \sum_{i=1}^{n}\left(100(x_i - x_{i-1}^2)^2 + (1 - x_i)^2\right).$$

We approach this problem by solving the TSSOS hierarchy with $r = 2$. Here we compare the results obtained with approximately smallest chordal extensions (**min**) and maximal chordal extensions (**max**). In Table 7.5, we can see that TSSOS scales much better with approximately smallest chordal extensions than with maximal chordal extensions while providing the same optimum.

Table 7.5: Results for the generalized Rosenbrock function.

n	min				max			
	s	mb	opt	time	s	mb	opt	time
20	1	21	18.25	0.19	1	58	18.25	8.2
					2	211	18.25	45
60	1	61	57.85	6.6	1	178	-	-
					2	1831	-	-
100	1	101	97.45	85	1	308	-	-
					2	5051	-	-
140	1	141	137.05	448	1	428	-	-
					2	9871	-	-
180	1	181	176.65	1495	1	548	-	-
					2	16291	-	-

7.7 Notes and sources

Besides the Newton polytope method and the approach given in Section 7.4, there are also other algorithms that provide a possibly smaller monomial basis; see for instance [KKW05] and [YC20].

The results on the TSSOS hierarchy presented in this chapter have been published in [WML21, WML21a]. The idea of exploiting TS in SOS decompositions was initially proposed in [WLX19] for the unconstrained case and sooner after extended to the constrained case in [WML21, WML21a]. The exploitation of sign symmetries in SOS decompositions was firstly discussed in [Lof09] in the unconstrained setting. Theorem 7.15, stated in [WML21, Theorem 6.5], relates the convergence of block structures arising from the TSSOS hierarchy (for a fixed relaxation order) to sign symmetries. For more extensive comparison of TSSOS with the polynomial optimization solvers GloptiPoly [HLL09], Yalmip [Lö04], and SparsePOP [WKK+08], the reader is referred to [WML21, WML21a].

Bibliography

[AM14]   Amir Ali Ahmadi and Anirudha Majumdar. Dsos and sd-
            sos optimization: Lp and socp-based alternatives to sum of
            squares optimization. In *2014 48th annual conference on infor-
            mation sciences and systems (CISS)*, pages 1–5. IEEE, 2014.

[ART03]   Erling D Andersen, Cornelis Roos, and Tamas Terlaky.
            On implementing a primal-dual interior-point method for
            conic quadratic optimization. *Mathematical Programming*,
            95(2):249–277, 2003.

[HLL09]   D. Henrion, Jean-Bernard Lasserre, and J. Löfberg. Glop-
            tiPoly 3: moments, optimization and semidefinite program-
            ming. *Optimization Methods and Software*, 24(4-5):pp. 761–779,
            August 2009.

[HP11]   Edward J Hancock and Antonis Papachristodoulou. Struc-
            tured sum of squares for networked systems analysis. In *2011
            50th IEEE Conference on Decision and Control and European Con-
            trol Conference*, pages 7236–7241. IEEE, 2011.

[KKW05]   Masakazu Kojima, Sunyoung Kim, and Hayato Waki. Spar-
            sity in sums of squares of polynomials. *Mathematical Program-
            ming*, 103(1):45–62, 2005.

[Las01]   Jean-Bernard Lasserre. Global Optimization with Polynomi-
            als and the Problem of Moments. *SIAM Journal on Optimiza-
            tion*, 11(3):796–817, 2001.

[Lof09]   Johan Lofberg. Pre-and post-processing sum-of-squares pro-
            grams in practice. *IEEE transactions on automatic control*,
            54(5):1007–1011, 2009.

[Lö04]   J. Löfberg. Yalmip : A toolbox for modeling and optimization
            in MATLAB. In *Proceedings of the CACSD Conference*, Taipei,
            Taiwan, 2004.

[MAT14]   Anirudha Majumdar, Amir Ali Ahmadi, and Russ Tedrake. Control and verification of high-dimensional systems with dsos and sdsos programming. In *53rd IEEE Conference on Decision and Control*, pages 394–401. IEEE, 2014.

[Meg10]   A Megretski. Systems polynomial optimization tools (spot). *Massachusetts Inst. Technol., Cambridge, MA, USA*, 2010.

[Rez78]   Bruce Reznick. Extremal psd forms with few terms. *Duke mathematical journal*, 45(2):363–374, 1978.

[WKK$^+$08] Hayato Waki, Sunyoung Kim, Masakazu Kojima, Masakazu Muramatsu, and Hiroshi Sugimoto. Algorithm 883: Sparsepop—a sparse semidefinite programming relaxation of polynomial optimization problems. *ACM Transactions on Mathematical Software (TOMS)*, 35(2):1–13, 2008.

[WLX19]   Jie Wang, Haokun Li, and Bican Xia. A new sparse sos decomposition algorithm based on term sparsity. In *Proceedings of the 2019 on International Symposium on Symbolic and Algebraic Computation*, pages 347–354, 2019.

[WML21a]   Jie Wang, Victor Magron, and Jean-Bernard Lasserre. Chordal-tssos: a moment-sos hierarchy that exploits term sparsity with chordal extension. *SIAM Journal on Optimization*, 31(1):114–141, 2021.

[WML21b]   Jie Wang, Victor Magron, and Jean-Bernard Lasserre. Tssos: A moment-sos hierarchy that exploits term sparsity. *SIAM Journal on Optimization*, 31(1):30–58, 2021.

[WMLM20] Jie Wang, Victor Magron, Jean B Lasserre, and Ngoc Hoang Anh Mai. Cs-tssos: Correlative and term sparsity for large-scale polynomial optimization. *arXiv preprint arXiv:2005.02828*, 2020.

[YC20]   Heng Yang and Luca Carlone. One ring to rule them all: Certifiably robust geometric perception with outliers. *arXiv preprint arXiv:2006.06769*, 2020.

Chapter 8

Exploiting both correlative and term sparsity

In previous chapters, we have seen that how CS or TS of POPs helps to reduce the size of SDP relaxations arising from the moment-SOS hierarchy. For large-scale POPs, it is natural to ask whether one can exploit CS and TS simultaneously to further reduce the size of SDP relaxations. As we shall see in this chapter, the answer is affirmative.

8.1 The CS-TSSOS hierarchy

Given a sparse POP, the underlying idea to exploit CS and TS simultaneously in the moment-SOS hierarchy consists of the following two steps:

(1) decomposing the set of variables into a tuple of cliques $\{I_k\}_{k \in [p]}$ by exploiting CS as in Chapter 3;

(2) applying the iterative procedure for exploiting TS to each decoupled subsystem involving variables $\mathbf{x}(I_k)$ as in Chapter 7.

More concretely, let us fix a relaxation order $r \geq r_{\min}$. Suppose that G^{csp} is the csp graph associated to POP (2.4) defined as in Chapter 3, $(G^{\mathrm{csp}})'$ is a chordal extension of G^{csp}, and $I_k, k \in [p]$ are the maximal cliques of $(G^{\mathrm{csp}})'$ with cardinality being denoted by $n_k, k \in [p]$. As in Chapter 3, the set of variables $\mathbf{x}$ is decomposed into $\mathbf{x}(I_1), \mathbf{x}(I_2), \ldots, \mathbf{x}(I_p)$ by exploiting CS. In addition, assume that the constraints are assigned to the variable cliques according to $J_1, \ldots, J_p, J'$ as defined in Chapter 3.

Now we apply the iterative procedure for exploiting TS to each subsystem involving variables $\mathbf{x}(I_k)$, $k \in [p]$ in the following way. Let

$$\mathscr{A} := \operatorname{supp}(f) \cup \bigcup_{j=1}^{m} \operatorname{supp}(g_j) \text{ and } \mathscr{A}_k := \{\alpha \in \mathscr{A} \mid \operatorname{supp}(\alpha) \subseteq I_k\} \quad (8.1)$$

for $k \in [p]$. Let $\mathbb{N}_{r-d_j}^{n_k}$ be the standard monomial basis for $j \in \{0\} \cup J_k, k \in [p]$. Let $G_{r,k}^{\mathrm{tsp}}$ be the tsp graph with nodes $\mathbb{N}_r^{n_k}$ associated to the support $\mathscr{A}_k$ defined as in Chapter 7, i.e., its node set is $\mathbb{N}_r^{n_k}$ and $\{\beta, \gamma\}$ is an edge if $\beta + \gamma \in \mathscr{A}_k \cup 2\mathbb{N}_r^{n_k}$. Note that here we embed $\mathbb{N}_r^{n_k}$ into $\mathbb{N}_r^{n}$ via the map $\alpha = (\alpha_i) \in \mathbb{N}_r^{n_k} \mapsto \alpha' = (\alpha_i') \in \mathbb{N}_r^{n}$ satisfying

$$\alpha_i' = \begin{cases} \alpha_i, & \text{if } i \in I_k, \\ 0, & \text{otherwise.} \end{cases}$$

Let us define $G_{r,k,0}^{(0)} := G_{r,k}^{\mathrm{tsp}}$ and $G_{r,k,j}^{(0)}, j \in J_k, k \in [p]$ are all empty graphs. Next for each $j \in \{0\} \cup J_k$ and each $k \in [p]$, we iteratively define an ascending chain of graphs $(G_{r,k,j}^{(s)}(V_{r,k,j}, E_{r,k,j}^{(s)}))_{s \geq 1}$ with $V_{r,k,j} := \mathbb{N}_{r-d_j}^{n_k}$ via two successive operations:

(1) **support extension.** Define $F_{r,k,j}^{(s)}$ to be the graph with nodes $V_{r,k,j}$ and with edges

$$E(F_{r,k,j}^{(s)}) = \left\{ \{\beta, \gamma\} \mid \beta \neq \gamma \in V_{r,k,j}, (\beta + \gamma + \operatorname{supp}(g_j)) \cap \mathscr{C}_r^{(s-1)} \neq \varnothing \right\}, \quad (8.2)$$

where

$$\mathscr{C}_r^{(s-1)} := \bigcup_{k=1}^{p} \left(\bigcup_{j \in \{0\} \cup J_k} \left(\operatorname{supp}(g_j) + \operatorname{supp}(G_{r,k,j}^{(s-1)}) \right) \right). \quad (8.3)$$

(2) **chordal extension.** Let

$$G_{r,k,j}^{(s)} := (F_{r,k,j}^{(s)})', \quad j \in \{0\} \cup J_k, k \in [p]. \quad (8.4)$$

It is clear by construction that the sequences of graphs $(G_{r,k,j}^{(s)})_{s \geq 1}$ stabilize for all $j \in \{0\} \cup J_k, k \in [p]$ after finitely many steps.

Example 8.1 *Let $f = 1 + x_1^2 + x_2^2 + x_3^2 + x_1 x_2 + x_2 x_3 + x_3$ and consider the unconstrained POP: $\min\{f(\mathbf{x}) : \mathbf{x} \in \mathbb{R}^n\}$. Take the relaxation order $r = r_{\min} = 1$. The variables are decomposed into two cliques: $\{x_1, x_2\}$ and $\{x_2, x_3\}$. The tsp graphs with respect to these two cliques are illustrated in Figure 8.1. The left graph corresponds to the first clique: x_1 and x_2 are connected because of the term*

x_1x_2. *The right graph corresponds to the second clique: 1 and x_3 are connected because of the term x_3; x_2 and x_3 are connected because of the term x_2x_3. It is not hard to see that the graph sequences $(G_{1,k}^{(s)})_{s\geq 1}, k = 1,2$ (the subscript j is omitted here since there is no constraint) stabilize at $s = 2$ if the maximal chordal extension is used in (8.4).*

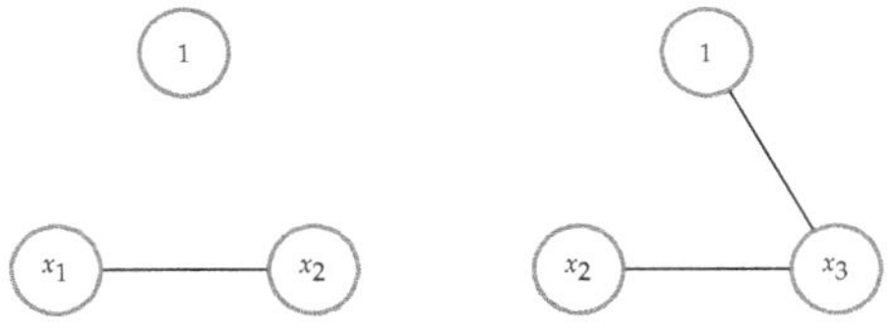

Figure 8.1: The tsp graphs of Example 8.1.

Let $t_{k,j} := |\mathbb{N}_{r-d_j}^{n_k}| = \binom{n_k+r-d_j}{r-d_j}$ for all k,j. Then with $s \geq 1$, the moment relaxation based on correlative-term sparsity for POP (2.4) is given by

$$
\mathbf{P}_{\text{cs-ts}}^{r,s} : \begin{cases}
\inf_{\mathbf{y}} & L_\mathbf{y}(f) \\
\text{s.t.} & \mathbf{B}_{G_{r,k,0}^{(s)}} \circ \mathbf{M}_r(\mathbf{y}, I_k) \in \Pi_{G_{r,k,0}^{(s)}}(\mathbf{S}_{t_{k,0}}^+), \quad k \in [p] \\
& \mathbf{B}_{G_{r,k,j}^{(s)}} \circ \mathbf{M}_{r-d_j}(g_j\mathbf{y}, I_k) \in \Pi_{G_{r,k,j}^{(s)}}(\mathbf{S}_{t_{k,j}}^+), \quad j \in J_k, k \in [p] \\
& L_\mathbf{y}(g_j) \geq 0, \quad j \in J' \\
& y_0 = 1
\end{cases}
$$

(8.5)

with optimum denoted by $f_{\text{cs-ts}}^{r,s}$.

For all k,j, let us write $\mathbf{M}_{r-d_j}(g_j\mathbf{y}, I_k) = \sum_\alpha \mathbf{D}_\alpha^{k,j} y_\alpha$ for appropriate symmetry matrices $\{\mathbf{D}_\alpha^{k,j}\}$ and $g_j = \sum_\alpha g_{j,\alpha} \mathbf{x}^\alpha$. Then for each $s \geq 1$, the dual of $\mathbf{P}_{\text{cs-ts}}^{r,s}$ (8.5) reads as

$$
(\mathbf{P}_{\text{cs-ts}}^{r,s})^* : \begin{cases}
\sup_{G_{k,j},\lambda_j,b} & b \\
\text{s.t.} & \sum_{k=1}^p \sum_{j\in\{0\}\cup J_k} \langle G_{k,j}, \mathbf{D}_\alpha^{k,j}\rangle + \sum_{j\in J'} \lambda_j g_{j,\alpha} \\
& \qquad\qquad +b\delta_{0\alpha} = f_\alpha, \quad \forall \alpha \in \mathscr{C}_r^{(s)} \\
& G_{k,j} \in \mathbf{S}_+^{t_{k,j}} \cap \mathbf{S}_{G_{r,k,j}^{(s)}}, \quad j \in \{0\} \cup J_k, k \in [p] \\
& \lambda_j \geq 0, \quad j \in J'
\end{cases}
$$

(8.6)

where $\mathscr{C}_r^{(s)}$ is defined in (8.3).

The primal-dual SDP relaxations (8.5)–(8.6) is called the CS-TS adpated moment-SOS (CS-TSSOS) hierarchy associated with **P** (2.4), which is indexed by two parameters: the relaxation order r and the sparse order s.

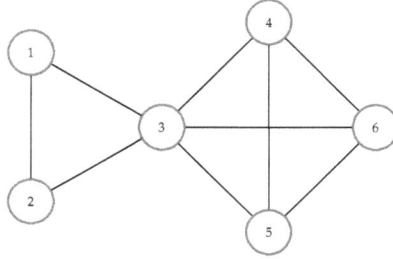

Figure 8.2: The csp graph of Example 8.2.

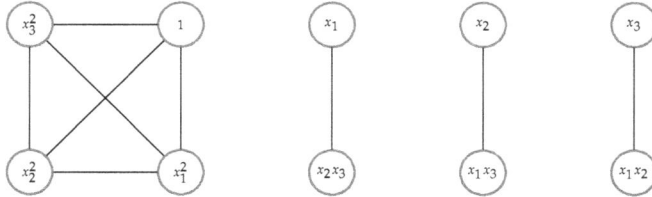

Figure 8.3: The tsp graph for the first clique of Example 8.2.

Example 8.2 Let $f = 1 + \sum_{i=1}^{6} x_i^4 + x_1 x_2 x_3 + x_3 x_4 x_5 + x_3 x_4 x_6 + x_3 x_5 x_6 + x_4 x_5 x_6$, and consider the unconstrained POP: $\min\{f(\mathbf{x}) : \mathbf{x} \in \mathbb{R}^n\}$. Let us apply the CS-TSSOS hierarchy (using the maximal chordal extension in (8.4)) to this problem by taking $r = r_{\min} = 2, s = 1$. First, according to the csp graph (Figure 8.2), we decompose the variables into two cliques: $\{x_1, x_2, x_3\}$ and $\{x_3, x_4, x_5, x_6\}$. The tsp graphs for the first clique and the second clique are shown in Figure 8.3 and Figure 8.4, respectively. For the first clique one obtains four blocks of SDP matrices with respective sizes $4, 2, 2, 2$. For the second clique one obtains two blocks of SDP matrices with respective sizes $5, 10$. As a result, the original SDP matrix of size 28 has been reduced to six blocks of maximal size 10.

Alternatively, if one applies the TSSOS hierarchy (using the maximal chordal extension in (7.15)) directly to this problem by taking $r = r_{\min} = 2, s = 1$ (i.e., without decomposing variables), then the tsp graph is shown in Figure 8.5 and one thereby obtains 11 PSD blocks with respective sizes $7, 2, 2, 2, 1, 1, 1, 1, 1, 1, 10$. Compared to the CS-TSSOS case, there are six additional blocks of size one and the two blocks with respective sizes $4, 5$ are replaced by a single block of size 7.

We summarize the basic properties of the CS-TSSOS hierarchy in the next theorem.

Theorem 8.3 Let $f \in \mathbb{R}[\mathbf{x}]$ and $\mathbf{X}$ be defined as in (2.1). Then the following hold:

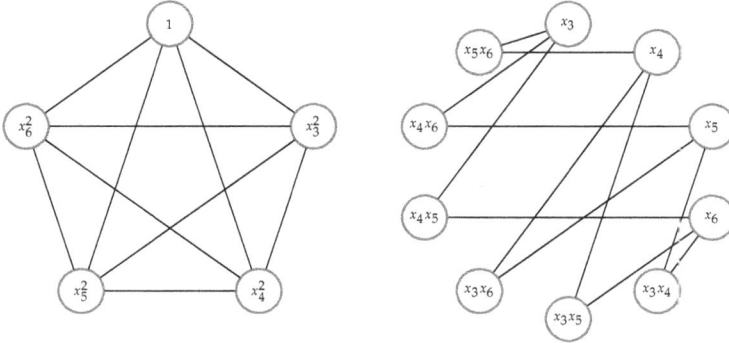

Figure 8.4: The tsp graph for the second clique of Example 8.2.

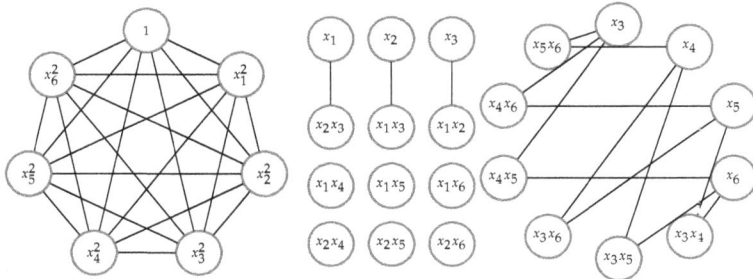

Figure 8.5: The tsp graph without decomposing variables of Example 8.2.

1. If $\mathbf{X}$ has a nonempty interior, then there is no duality gap between $\mathbf{P}^{r,s}_{\text{cs-ts}}$ and $(\mathbf{P}^{r,s}_{\text{cs-ts}})^*$ for any $r \geq r_{\min}$ and $s \geq 1$.

2. For any $r \geq r_{\min}$, the sequence $(f^{r,s}_{\text{cs-ts}})_{s \geq 1}$ is monotonically non-decreasing and $f^{r,s}_{\text{cs-ts}} \leq f^r_{\text{cs}}$ for all s with f^r_{cs} being defined in Section 3.3.

3. For any $s \geq 1$, the sequence $(f^{r,s}_{\text{cs-ts}})_{r \geq r_{\min}}$ is monotonically non-decreasing.

PROOF 1. By the duality theory of convex programming, this easily follows from Theorem 3.6 of [Las06] and Theorem 1.4.

2. By construction, we have $G^{(s)}_{r,k,j} \subseteq G^{(s+1)}_{r,k,j}$ for all r, k, j and for all s. It follows that each maximal clique of $G^{(s)}_{r,k,j}$ is contained in some maximal

clique of $G_{r,k,j}^{(s+1)}$. Hence by Theorem 1.4, $\mathbf{P}_{\text{cs-ts}}^{r,s}$ is a relaxation of $\mathbf{P}_{\text{cs-ts}}^{r,s+1}$ and is clearly also a relaxation of $\mathbf{P}_{\text{cs}}^{r}$. Therefore, $(f_{\text{cs-ts}}^{r,s})_{s \geq 1}$ is monotonically nondecreasing and $f_{\text{cs-ts}}^{r,s} \leq f_{\text{cs}}^{r}$ for all s.

3. The conclusion follows if we can show that the inclusion $G_{r,k,j}^{(s)} \subseteq G_{r+1,k,j}^{(s)}$ holds for all r, k, j, s, since by Theorem 1.4 this implies that $\mathbf{P}_{\text{cs-ts}}^{r,s}$ is a relaxation of $\mathbf{P}_{\text{cs-ts}}^{r+1,s}$. Let us prove $G_{r,k,j}^{(s)} \subseteq G_{r+1,k,j}^{(s)}$ by induction on s. For $s = 1$, we have $G_{r,k,0}^{(0)} = G_{r,k}^{\text{tsp}} \subseteq G_{r+1,k}^{\text{tsp}} = G_{r+1,k,0}^{(0)}$, which together with (8.2)–(8.3) implies that $F_{r,k,j}^{(1)} \subseteq F_{r+1,k,j}^{(1)}$ for $j \in \{0\} \cup J_k, k \in [p]$. It then follows that $G_{r,k,j}^{(1)} = (F_{r,k,j}^{(1)})' \subseteq (F_{r+1,k,j}^{(1)})' = G_{r+1,k,j}^{(1)}$. Now assume that $G_{r,k,j}^{(s)} \subseteq G_{r+1,k,j}^{(s)}, j \in \{0\} \cup J_k, k \in [p]$, hold for some $s \geq 1$. Then by (8.2)–(8.3) and by the induction hypothesis, we have $F_{r,k,j}^{(s+1)} \subseteq F_{r+1,k,j}^{(s+1)}$ for $j \in \{0\} \cup J_k, k \in [p]$. Thus $G_{r,k,j}^{(s+1)} = (F_{r,k,j}^{(s+1)})' \subseteq (F_{r+1,k,j}^{(s+1)})' = G_{r+1,k,j}^{(s+1)}$ which completes the induction. □

From Theorem 8.3, we deduce the following two-level hierarchy of lower bounds for the optimum $f_{\min}$ of $\mathbf{P}$ (2.4):

$$
\begin{array}{ccccccc}
f_{\text{cs-ts}}^{r_{\min},1} & \leq & f_{\text{cs-ts}}^{r_{\min},2} & \leq & \cdots & \leq & f_{\text{cs-ts}}^{r_{\min}} \\
\wedge| & & \wedge| & & & & \wedge| \\
f_{\text{cs-ts}}^{r_{\min}+1,1} & \leq & f_{\text{cs-ts}}^{r_{\min}+1,2} & \leq & \cdots & \leq & f_{\text{cs-ts}}^{r_{\min}+1} \\
\wedge| & & \wedge| & & & & \wedge| \\
\vdots & & \vdots & & \vdots & & \vdots \\
\wedge| & & \wedge| & & & & \wedge| \\
f_{\text{cs-ts}}^{r,1} & \leq & f_{\text{cs-ts}}^{r,2} & \leq & \cdots & \leq & f_{\text{cs-ts}}^{r} \\
\wedge| & & \wedge| & & & & \wedge| \\
\vdots & & \vdots & & \vdots & & \vdots
\end{array}
\tag{8.7}
$$

As we have known for the TSSOS hierarchy, the block structure arising from the CS-TSSOS hierarchy is consistent with the sign symmetries of the POP. More precisely, we have the following theorem.

Theorem 8.4 *Let $\mathscr{A}$ be defined as in (8.1), $\mathscr{C}_r^{(s)}$ be defined as in (8.3), and assume that the sign symmetries of $\mathscr{A}$ are represented by the column vectors of the binary matrix $\mathbf{R}$. Then for any $r \geq r_{\min}, s \geq 1$ and any $\alpha \in \mathscr{C}_r^{(s)}$, it holds $\mathbf{R}^{\mathsf{T}} \alpha \equiv \mathbf{0} \pmod 2$. As a consequence, if β, γ belong to the same block in the CS-TSSOS relaxations, then $\mathbf{R}^{\mathsf{T}}(\beta + \gamma) \equiv \mathbf{0} \pmod 2$.*

8.2   Global convergence

We next show that if the chordal extension in (8.5) is chosen to be *maximal*, then for any relaxation order $r \geq r_{\min}$, the sequence of optima $(f_{\text{cs-ts}}^{r,s})_{s \geq 1}$ arising from the CS-TSSOS hierarchy converges to the optimum f_{cs}^r of the CSSOS relaxation.

It is clear by construction that the sequences of graphs $(G_{r,k,j}^{(s)})_{s \geq 1}$ stabilize for all $j \in \{0\} \cup J_k, k \in [p]$ after finitely many steps. We denote the resulting stabilized graphs by $G_{r,k,j}^{(\bullet)}, j \in \{0\} \cup J_k, k \in [p]$ and the corresponding SDP (8.5) by $\mathbf{P}_{\text{cs-ts}}^{r,\bullet}$.

Theorem 8.5 *If one uses the maximal chordal extension in (8.4), then for any $r \geq r_{\min}$, the sequence $(f_{\text{cs-ts}}^{r,s})_{s \geq 1}$ converges to f_{cs}^r in finitely many steps.*

PROOF   Let $\mathbf{y} = (y_\alpha)$ be an arbitrary feasible solution of $\mathbf{P}_{\text{cs-ts}}^{r,\bullet}$ and $f_{\text{cs-ts}}^{r,\bullet}$ be the optimum of $\mathbf{P}_{\text{cs-ts}}^{r,\bullet}$. Note that $\{y_\alpha \mid \alpha \in \bigcup_{k=1}^p (\bigcup_{j \in \{0\} \cup J_k} (\text{supp}(g_j) + \text{supp}(G_{r,k,j}^{(\bullet)})))\}$ is the set of decision variables involved in $\mathbf{P}_{\text{cs-ts}}^{r,\bullet}$. Let $\mathscr{R}$ be the set of decision variables involved in $\mathbf{P}_{\text{cs}}^r$ (3.6). We then define a vector $\bar{\mathbf{y}} = (\bar{y}_\alpha)_{\alpha \in \mathscr{R}}$ as follows:

$$\bar{y}_\alpha = \begin{cases} y_\alpha, & \text{if } \alpha \in \bigcup_{k=1}^p (\bigcup_{j \in \{0\} \cup J_k} (\text{supp}(g_j) + \text{supp}(G_{r,k,j}^{(\bullet)}))), \\ 0, & \text{otherwise.} \end{cases}$$

By construction and since $G_{r,k,j}^{(\bullet)}$ stabilizes under support extension for all k, j, we have $\mathbf{M}_{r-d_j}(g_j \bar{\mathbf{y}}, I_k) = \mathbf{B}_{G_{r,k,j}^{(\bullet)}} \circ \mathbf{M}_{r-d_j}(g_j \mathbf{y}, I_k)$. As the maximal chordal extension is chosen for (8.4), the matrix $\mathbf{B}_{G_{r,k,j}^{(\bullet)}} \circ \mathbf{M}_{r-d_j}(g_j \mathbf{y}, I_k)$ is block diagonal up to permutation. It follows from $\mathbf{B}_{G_{r,k,j}^{(\bullet)}} \circ \mathbf{M}_{r-d_j}(g_j \mathbf{y}, I_k) \in \Pi_{G_{r,k,j}^{(\bullet)}}(\mathbf{S}_+^{t_{k,j}})$ that $\mathbf{M}_{r-d_j}(g_j \bar{\mathbf{y}}, I_k) \succeq 0$ for $j \in \{0\} \cup J_k, k \in [p]$. Therefore $\bar{\mathbf{y}}$ is a feasible solution of $\mathbf{P}_{\text{cs}}^r$ and so $L_{\mathbf{y}}(f) = L_{\bar{\mathbf{y}}}(f) \geq f_{\text{cs}}^r$. Hence $f_{\text{cs-ts}}^{r,\bullet} \geq f_{\text{cs}}^r$ since $\mathbf{y}$ is an arbitrary feasible solution of $\mathbf{P}_{\text{cs-ts}}^{r,\bullet}$. By Theorem 8.3, we already have $f_{\text{cs-ts}}^{r,\bullet} \leq f_{\text{cs}}^r$. Therefore, $f_{\text{cs-ts}}^{r,\bullet} = f_{\text{cs}}^r$.   □

By Theorem 3.6 in [Las06], the sequence $(f_{\text{cs}}^r)_{r \geq r_{\min}}$ converges to the global optimum $f_{\min}$ of POP (2.4) (after adding some redundant quadratic constraints). Therefore, this together with Theorem 8.5 offers the global convergence of the CS-TSSOS hierarchy.

Proceeding along Theorem 8.3, we are able to provide a *sparse representation* theorem based on both CS and TS for a polynomial positive on a compact basic semialgebraic set.

Theorem 8.6 *Let $f \in \mathbb{R}[\mathbf{x}]$, $\mathbf{X} \subseteq \mathbb{R}^n$ and $\{I_k\}_{k=1}^p, \{J_k\}_{k=1}^p$ be defined in Assumption 3.1. Assume that the sign symmetries of $\mathscr{A} = \mathrm{supp}(f) \cup \bigcup_{j=1}^m \mathrm{supp}(g_j)$ are represented by the columns of the binary matrix $\mathbf{R}$. If f is positive on $\mathbf{X}$, then f admits a representation of form*

$$f = \sum_{k=1}^p \left(\sigma_{k,0} + \sum_{j \in J_k} \sigma_{k,j} g_j \right), \qquad (8.8)$$

for some polynomials $\sigma_{k,j} \in \Sigma[\mathbf{x}(I_k)], j \in \{0\} \cup J_k, k \in [p]$, satisfying $\mathbf{R}^\mathsf{T} \alpha \equiv \mathbf{0} \ (\mathrm{mod}\ 2)$ for any $\alpha \in \mathrm{supp}(\sigma_{k,j})$.

PROOF By Corollary 3.9 of [Las06] (or Theorem 5 of [GNS07]), there exist polynomials $\sigma'_{k,j} \in \Sigma[\mathbf{x}(I_k)], j \in \{0\} \cup J_k, k \in [p]$ such that

$$f = \sum_{k=1}^p \left(\sigma'_{k,0} + \sum_{j \in J_k} \sigma'_{k,j} g_j \right). \qquad (8.9)$$

Let $r = \max\{\lceil \deg(\sigma'_{k,j} g_j)/2 \rceil : j \in \{0\} \cup J_k, k \in [p]\}$. Let $\mathbf{G}'_{k,j}$ be a PSD Gram matrix associated with $\sigma'_{k,j}$ and indexed by the monomial basis $\mathbb{N}^{n_k}_{r-d_j}$. Then for all k, j, we define $\mathbf{G}_{k,j} \in \mathbf{S}^{t_{k,j}}$ (indexed by $\mathbb{N}^{n_k}_{r-d_j}$) by

$$[\mathbf{G}_{k,j}]_{\beta\gamma} := \begin{cases} [Q'_{k,j}]_{\beta\gamma}, & \text{if } \mathbf{R}^\mathsf{T}(\beta + \gamma) \equiv \mathbf{0} \ (\mathrm{mod}\ 2), \\ 0, & \text{otherwise,} \end{cases}$$

and let $\sigma_{k,j} = (\mathbf{x}^{\mathbb{N}^{n_k}_{r-d_j}})^\mathsf{T} \mathbf{G}_{k,j} \mathbf{x}^{\mathbb{N}^{n_k}_{r-d_j}}$. One can easily verify that $\mathbf{G}_{k,j}$ is block diagonal up to permutation (see also [WML21]) and each block is a principal submatrix of $\mathbf{G}'_{k,j}$. Then the positive semidefiniteness of $\mathbf{G}'_{k,j}$ implies that $\mathbf{G}_{k,j}$ is also positive semidefinite. Thus $\sigma_{k,j} \in \Sigma[\mathbf{x}(I_k)]$.

By construction, substituting $\sigma'_{k,j}$ with $\sigma_{k,j}$ in (8.9) boils down to removing the terms with exponents α that do not satisfy $\mathbf{R}^\mathsf{T} \alpha \equiv \mathbf{0} \ (\mathrm{mod}\ 2)$ from the right-hand side of (8.9). Since any $\alpha \in \mathrm{supp}(f)$ satisfies $\mathbf{R}^\mathsf{T} \alpha \equiv \mathbf{0}$ (mod 2), this does not change the match of coefficients on both sides of the equality. Thus we obtain

$$f = \sum_{k=1}^{p} \left(\sigma_{k,0} + \sum_{j \in J_k} \sigma_{k,j} g_j \right)$$

with the desired property.   □

8.3 Extracting a solution

In the case of the dense moment-SOS hierarchy, there is a standard procedure described in [HL05] to extract globally optimal solutions when the moment matrix satisfies the so-called flatness condition. This procedure was generalized to the correlative sparse setting in [Las06, §3.3] and [ND09]. In the term sparse setting, however, the corresponding procedure cannot be applied because the information on the moment matrix is incomplete. In order to extract a solution in this case, we may add an order-one (dense) moment matrix for each clique in (8.5):

$$\begin{cases} \inf_{\mathbf{y}} & L_{\mathbf{y}}(f) \\ \text{s.t.} & \mathbf{M}_r(\mathbf{y}, I_k) \in \Pi_{G_{r,k,0}^{(s)}}(\mathbf{S}_{t_{k,0}}^+), \quad k \in [p] \\ & \mathbf{M}_1(\mathbf{y}, I_k) \succeq 0, \quad k \in [p] \\ & \mathbf{M}_{r-d_j}(g_j \mathbf{y}, I_k) \in \Pi_{G_{r,k,j}^{(s)}}(\mathbf{S}_{t_{k,j}}^+), \quad j \in J_k, \quad k \in [p] \\ & L_{\mathbf{y}}(g_j) \geq 0, \quad j \in J' \\ & y_0 = 1. \end{cases} \quad (8.10)$$

Let $\mathbf{y}^{\text{opt}}$ be an optimal solution of (8.10). Typically, $\mathbf{M}_1(\mathbf{y}^{\text{opt}}, I_k)$ (after identifying sufficiently small entries with zeros) is a block diagonal matrix (up to permutation). If for all k every block of $\mathbf{M}_1(\mathbf{y}^{\text{opt}}, I_k)$ is of rank one, then a globally optimal solution $\mathbf{x}^{\text{opt}}$ to $\mathbf{P}$ (2.4) which is unique up to sign symmetries can be extracted ([Las06, Theorem 3.7]), and the global optimality is certified. Otherwise, the relaxation might be not exact or yield multiple global solutions.

Remark 8.7 *Note that (8.10) is a tighter relaxation of* $\mathbf{P}$ *(2.4) than* $\mathbf{P}_{\text{cs-ts}}^{r,s}$ *(8 5), and so might provide a better lower bound for* $\mathbf{P}$. *In particular, if* $\mathbf{P}$ *is a QCQP, then (8.10) is always tighter than Shor's relaxation of* $\mathbf{P}$.

8.4 A minimal initial relaxation step

For POP (2.4), suppose that f is not a homogeneous polynomial or the constraint polynomials $\{g_j\}_{j \in [m]}$ are of different degrees. Then instead of

using the uniform minimum relaxation order $r_{\min}$, it might be more beneficial, from the computational point of view, to assign different relaxation
orders to different subsystems obtained from the csp for the initial relaxation step of the CS-TSSOS hierarchy. To this end, we redefine the csp
graph $G^{\mathrm{icsp}}(V, E)$ as follows: $V = [n]$ and $\{i, j\} \in E$ whenever there exists
$\alpha \in \mathscr{A}$ such that $\{i, j\} \subseteq \mathrm{supp}(\alpha)$. This is clearly a subgraph of G^{csp} defined in Chapter 3 and hence typically admits a smaller chordal extension.
Let $(G^{\mathrm{icsp}})'$ be a chordal extension of G^{icsp} and $\{I_k\}_{k \in [p]}$ be the list of max-
imal cliques of $(G^{\mathrm{icsp}})'$ with $n_k := |I_k|$. Now we partition the constraint
polynomials $\{g_j\}_{j \in [m]}$ into groups $\{g_j \mid j \in J_k\}_{k \in [p]}$ and $\{g_j \mid j \in J'\}$ which
satisfy

(1) $J_1, \ldots, J_p, J' \subseteq [m]$ are pairwise disjoint and $\bigcup_{k=1}^{p} J_k \cup J' = [m]$;

(2) for any $j \in J_k$, $\bigcup_{\alpha \in \mathrm{supp}(g_j)} \mathrm{supp}(\alpha) \subseteq I_k, k \in [p]$;

(3) for any $j \in J'$, $\bigcup_{\alpha \in \mathrm{supp}(g_j)} \mathrm{supp}(\alpha) \not\subseteq I_k$ for all $k \in [p]$.

Suppose f decomposes as $f = \sum_{k \in [p]} f_k$ such that $\bigcup_{\alpha \in \mathrm{supp}(f_k)} \mathrm{supp}(\alpha) \subseteq$
I_k for $k \in [p]$. We define the vector of minimum relaxation orders $\mathbf{o} =$
$(o_k)_k \in \mathbb{N}^p$ with $o_k := \max\{\{d_j : j \in J_k\} \cup \{\lceil \deg(f_k)/2 \rceil\}\}$. Then with $s \geq$
1, we define the following minimal initial relaxation step of the CS-TSSOS
hierarchy:

$$\begin{cases} \inf_{\mathbf{y}} \quad L_{\mathbf{y}}(f) \\ \text{s.t.} \quad \mathbf{B}_{G^{(s)}_{o_k,k,0}} \circ \mathbf{M}_{o_k}(\mathbf{y}, I_k) \in \Pi_{G^{(s)}_{o_k,k,0}}(\mathbf{S}_+^{t_{k,0}}), \quad k \in [p] \\ \qquad \mathbf{B}_{G^{(s)}_{o_k,k,j}} \circ \mathbf{M}_{o_k - d_j}(g_j \mathbf{y}, I_k) \in \Pi_{G^{(s)}_{o_k,k,j}}(\mathbf{S}_+^{t_{k,j}}), \quad j \in J_k, k \in [p] \\ \qquad L_{\mathbf{y}}(g_j) \geq 0, \quad j \in J' \\ \qquad y_0 = 1 \end{cases} \qquad (8.11)$$

where $G^{(s)}_{o_k,k,j}, j \in J_k, k \in [p]$ are defined in the same spirit with Section 8.1
and $t_{k,j} := \binom{n_k + o_k - d_j}{o_k - d_j}$ for all k, j.

8.5 Numerical experiments

In this section, we report some numerical results of the proposed CS-TSSOS
hierarchy. All numerical examples were computed on an Intel Core i5-
8265U@1.60GHz CPU with 8GB RAM memory.

8.5.1 Benchmarks for constrained POPs

Consider the following POP:

$$
\begin{cases}
\inf\limits_{\mathbf{x}} & f_{\mathrm{gR}} \quad (\text{resp. } f_{\mathrm{Bt}} \text{ or } f_{\mathrm{cW}}) \\
\text{s.t.} & 1 - \left(\sum_{i=20j-19}^{20j} x_i^2\right) \geq 0, \quad j = 1, 2, \ldots, n/20
\end{cases}
\tag{8.12}
$$

with $20|n$, where the objective function is respectively given by
• the generalized Rosenbrock function

$$
f_{\mathrm{gR}}(\mathbf{x}) = 1 + \sum_{i=2}^{n} (100(x_i - x_{i-1}^2)^2 + (1 - x_i)^2),
$$

• the Broyden tridiagonal function

$$
f_{\mathrm{Bt}}(\mathbf{x}) = ((3 - 2x_1)x_1 - 2x_2 + 1)^2 + \sum_{i=2}^{n-1} ((3 - 2x_i)x_i - x_{i-1} - 2x_{i+1} + 1)^2
$$
$$
+ ((3 - 2x_n)x_n - x_{n-1} + 1)^2,
$$

• the chained Wood function

$$
f_{\mathrm{cW}}(\mathbf{x}) = 1 + \sum_{i \in J} (100(x_{i+1} - x_i^2)^2 + (1 - x_i)^2 + 90(x_{i+3} - x_{i+2}^2)^2
$$
$$
+ (1 - x_{i+2})^2 + 10(x_{i+1} + x_{i+3} - 2)^2 + 0.1(x_{i+1} - x_{i+3})^2),
$$

where $J = \{1, 3, 5, \ldots, n - 3\}$ and $4|n$.

We solve the CSSOS relaxation with $r = 2$, the TSSOS relaxation with $r = 2, s = 1$, and the CS-TSSOS relaxation with $r = 2, s = 1$, where approximately smallest chordal extensions are used for TS. The results are presented in Tables 8.1–8.3, in which "mb" denotes the maximal size of PSD blocks involved in the relaxations, "opt" denotes the optimum, "time" denotes running time in seconds, and "-" indicates an out of memory error. We see that CSSOS, TSSOS and CS-TSSOS yield almost the same optimum while CS-TSSOS is the most efficient and scalable approach among them.

8.5.2 The Max-Cut problem

The Max-Cut problem is one of the basic combinatorial optimization problems, which is known to be NP-hard. Let $G(V, E)$ be an undirected graph with $V = \{1, \ldots, n\}$ and with edge weights w_{ij} for $\{i, j\} \in E$. Then the Max-Cut problem for G can be formulated as a QCQP in binary variables:

$$
\begin{cases}
\max\limits_{\mathbf{x}} & \frac{1}{2}\sum_{\{i,j\} \in E} w_{ij}(1 - x_i x_j) \\
\text{s.t.} & 1 - x_i^2 = 0, \quad i = 1, \ldots, n.
\end{cases}
\tag{8.13}
$$

Table 8.1: Results for the generalized Rosenbrock function ($r = 2$).

n	CSSOS			TSSOS			CS-TSSOS		
	mb	opt	time	mb	opt	time	mb	opt	time
100	231	97.445	377	101	97.436	31.3	21	97.436	0.54
200	231	-	-	201	196.41	1327	21	196.41	1.27
300	231	-	-	-	-	-	21	295.39	2.26
400	231	-	-	-	-	-	21	394.37	3.36
500	231	-	-	-	-	-	21	493.35	4.65
1000	231	-	-	-	-	-	21	988.24	15.8

Table 8.2: Results for the Broyden tridiagonal function ($r = 2$).

n	CSSOS			TSSOS			CS-TSSOS		
	mb	opt	time	mb	opt	time	mb	opt	time
100	231	79.834	519	103	79.834	104	23	79.834	1.96
200	231	-	-	-	-	-	23	160.83	4.88
300	231	-	-	-	-	-	23	241.83	8.67
400	231	-	-	-	-	-	23	322.83	13.3
500	231	-	-	-	-	-	23	403.83	19.9
1000	231	-	-	-	-	-	23	808.83	57.5

Table 8.3: Results for the chained Wood function ($r = 2$).

n	CSSOS			TSSOS			CS-TSSOS		
	mb	opt	time	mb	opt	time	mb	opt	time
100	231	1485.8	505	101	1485.8	43.2	21	1485.8	0.73
200	231	-	-	201	3004.5	1238	21	3004.5	1.91
300	231	-	-	-	-	-	21	4523.6	3.39
400	231	-	-	-	-	-	21	6042.0	5.72
500	231	-	-	-	-	-	21	7560.7	7.88
1000	231	-	-	-	-	-	21	15155	23.0

For the numerical experiments, we build random Max-Cut instances with a block-band sparsity pattern (illustrated in Figure 8.6) which consists of l blocks of size 25 and two bands of width 5. There are ten such Max-Cut instances with $l = 20, 40, 60, 80, 100, 120, 140, 160, 180, 200$, respectively.[1]

---

[1]The instances are available at https://wangjie212.github.io/jiewang/code.html.

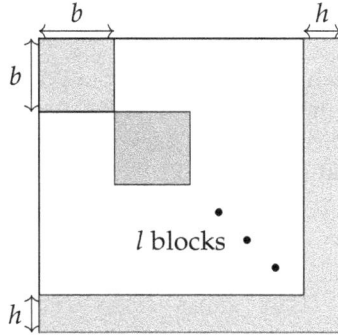

Figure 8.6: The block-band sparsity pattern. l: the number of blocks, b: the size of blocks, h: the width of bands.

For each instance, we solve Shor's relaxation, the CSSOS hierarchy with $r = 2$, and the CS-TSSOS hierarchy with $r = 2, s = 1$, where the maximal chordal extension is used for TS. The results are reported in Table 8.4. We see that for each instance, both CSSOS and CS-TSSOS signif-icantly improve the bound obtained from Shor's relaxation. Meanwhile, CS-TSSOS is several times faster than CSSOS at the cost of possibly pro-viding a sightly weaker bound.

Table 8.4: Results for Max-Cut instances. Only integer parts of optima are preserved. "mc" denotes the maximal size of variable cliques.

name	node	edge	mc	Shor	CSSOS			CS-TSSOS		
				opt	mb	opt	time	mb	opt	time
g20	505	2045	14	570	120	488	51.2	92	488	19.6
g40	1005	3441	14	1032	120	885	134	92	893	41.1
g60	1505	4874	14	1439	120	1227	183	92	1247	71.3
g80	2005	6035	15	1899	136	1638	167	106	1669	84.8
g100	2505	7320	14	2398	120	2073	262	92	2128	112
g120	3005	8431	14	2731	120	2358	221	79	2443	127
g140	3505	9658	13	3115	105	2701	250	79	2812	153
g160	4005	10677	14	3670	120	3202	294	79	3404	166
g180	4505	12081	13	4054	105	3525	354	79	3666	246
g200	5005	13240	13	4584	105	4003	374	79	4218	262

8.6 Notes and sources

The material from this chapter is issued from [WMLM20]. A proof of Theorem 8.4 can be found in Section 3.2 of [WMLM20].

Newton and Papachristodoulou have used the CS-TSSOS hierarchy for neural network verification in [NP22].

Bibliography

[GNS07]   David Grimm, Tim Netzer, and Markus Schweighofer. A note
            on the representation of positive polynomials with structured
            sparsity. *Archiv der Mathematik*, 89(5):399–403, 2007.

[HL05]   D. Henrion and Jean-Bernard Lasserre. *Detecting Global Op-
            timality and Extracting Solutions in GloptiPoly*, pages 293–310.
            Springer Berlin Heidelberg, Berlin, Heidelberg, 2005.

[Las06]   Jean B Lasserre. Convergent sdp-relaxations in polynomial
            optimization with sparsity. *SIAM Journal on Optimization*,
            17(3):822–843, 2006.

[ND09]   Jiawang Nie and James Demmel. Sparse sos relaxations for
            minimizing functions that are summations of small polyno-
            mials. *SIAM Journal on Optimization*, 19(4):1534–1558, 2009.

[NP22]   Matthew Newton and Antonis Papachristodoulou. Sparse
            polynomial optimisation for neural network verification.
            arXiv preprint arXiv:2202.02241, 2022.

[WML21]   Jie Wang, Victor Magron, and Jean-Bernard Lasserre. Tssos:
            A moment-sos hierarchy that exploits term sparsity. *SIAM
            Journal on Optimization*, 31(1):30–58, 2021.

[WMLM20]   Jie Wang, Victor Magron, Jean B Lasserre, and Ngoc
            Hoang Anh Mai. Cs-tssos: Correlative and term spar-
            sity for large-scale polynomial optimization. *arXiv preprint
            arXiv:2005.02828*, 2020.

Chapter 9

Application in optimal power flow

In this chapter, we apply the CS-TSSOS hierarchy to the famous alternating current optimal power flow (AC-OPF) problem, which can be formulized as a POP either with real variables [BEGL20, GMM15] or with complex variables [JM18]. To tackle POPs in complex variables, we first provide ingredients for extending the moment-SOS hierarchy to the complex case.

9.1 Extension to complex polynomial optimization

We start by introducing some notations. Let $\mathbf{z} = (z_1, \ldots, z_n)$ be a tuple of complex variables and $\bar{\mathbf{z}} = (\bar{z}_1, \ldots, \bar{z}_n)$ be its conjugate. We denote by $\mathbb{C}[\mathbf{z}] := \mathbb{C}[z_1, \ldots, z_n]$, $\mathbb{C}[\mathbf{z}, \bar{\mathbf{z}}] := \mathbb{C}[z_1, \ldots, z_n, \bar{z}_1, \ldots, \bar{z}_n]$ the complex polynomial ring in $\mathbf{z}$, the complex polynomial ring in $\mathbf{z}, \bar{\mathbf{z}}$, respectively. A polynomial $f \in \mathbb{C}[\mathbf{z}, \bar{\mathbf{z}}]$ can be written as $f = \sum_{(\beta,\gamma) \in \mathscr{A}} f_{\beta,\gamma} \mathbf{z}^\beta \bar{\mathbf{z}}^\gamma$ with $\mathscr{A} \subseteq \mathbb{N}^n \times \mathbb{N}^n$ and $f_{\beta,\gamma} \in \mathbb{C}, \mathbf{z}^\beta = z_1^{\beta_1} \cdots z_n^{\beta_n}, \bar{\mathbf{z}}^\gamma = \bar{z}_1^{\gamma_1} \cdots \bar{z}_n^{\gamma_n}$. The *support* of f is defined by $\operatorname{supp}(f) = \{(\beta, \gamma) \in \mathscr{A} \mid f_{\beta,\gamma} \neq 0\}$. The *conjugate* of f is $\bar{f} = \sum_{(\beta,\gamma) \in \mathscr{A}} \bar{f}_{\beta,\gamma} \mathbf{z}^\gamma \bar{\mathbf{z}}^\beta$. A polynomial $\sigma = \sum_{(\beta,\gamma)} \sigma_{\beta,\gamma} \mathbf{z}^\beta \bar{\mathbf{z}}^\gamma \in \mathbb{C}[\mathbf{z}, \bar{\mathbf{z}}]$ is called an *Hermitian sum of squares (HSOS)* if there exist polynomials $f_i \in \mathbb{C}[\mathbf{z}], i \in [t]$ such that $\sigma = \sum_{i=1}^t f_i \bar{f}_i$. For a positive integer m, the set of $m \times m$ Hermitian matrices is denoted by $\mathbb{H}_m$ and the set of $m \times m$ PSD Hermitian matrices is denoted by $\mathbb{H}_m^+$.

A complex polynomial optimization problem (CPOP) is given by

$$
\begin{cases}
\inf_{\mathbf{z} \in \mathbb{C}^n} & f(\mathbf{z}, \bar{\mathbf{z}}) := \sum_{\alpha,\beta} f_{\alpha,\beta} \mathbf{z}^\alpha \bar{\mathbf{z}}^\beta \\
\text{s.t.} & g_j(\mathbf{z}, \bar{\mathbf{z}}) := \sum_{\alpha,\beta} g_{j,\alpha,\beta} \mathbf{z}^\alpha \bar{\mathbf{z}}^\beta \geq 0, \quad j \in [m]
\end{cases}
\tag{9.1}
$$

where the functions $f, g_1, \ldots, g_m$ are real-valued polynomials and their co-efficients satisfy $f_{\alpha,\beta} = \bar{f}_{\beta,\alpha}$, and $g_{j,\alpha,\beta} = \bar{g}_{j,\beta,\alpha}$. There are two ways to construct a "moment-SOS" hierarchy for CPOP (9.1). The first way is in-troducing real variables for both real and imaginary parts of each complex variable in (9.1), i.e., letting $z_i = x_i + x_{i+n}\mathbf{i}$ for $i \in [n]$. Then one can convert CPOP (9.1) to a POP involving only real variables at the price of doubling the number of variables. Hence the usual real moment-SOS hier-archy applies to the resulting real POP. On the other hand, as the second way, it might be advantageous to handle CPOP (9.1) directly with the com-plex moment-HSOS hierarchy introduced in [JM18]. To this end, we define the *complex* moment matrix $\mathbf{M}_r^c(\mathbf{y})$ ($r \in \mathbb{N}$) by

$$[\mathbf{M}_r^c(\mathbf{y})]_{\beta,\gamma} := L_{\mathbf{y}}^c(\mathbf{z}^\beta \bar{\mathbf{z}}^\gamma) = y_{\beta,\gamma}, \quad \forall \beta, \gamma \in \mathbb{N}_r^n,$$

where $\mathbf{y} = (y_{\beta,\gamma})_{(\beta,\gamma) \in \mathbb{N}^n \times \mathbb{N}^n} \subseteq \mathbb{C}$ is a sequence indexed by $(\beta, \gamma) \in \mathbb{N}^n \times \mathbb{N}^n$ satisfying $y_{\beta,\gamma} = \bar{y}_{\gamma,\beta}$, and $L_{\mathbf{y}}^c : \mathbb{C}[\mathbf{z}, \bar{\mathbf{z}}] \to \mathbb{R}$ is the linear functional such that

$$f = \sum_{(\beta,\gamma)} f_{\beta,\gamma} \mathbf{z}^\beta \bar{\mathbf{z}}^\gamma \mapsto L_{\mathbf{y}}^c(f) = \sum_{(\beta,\gamma)} f_{\beta,\gamma} y_{\beta,\gamma}.$$

Suppose that $g = \sum_{(\beta',\gamma')} g_{\beta',\gamma'} \mathbf{z}^{\beta'} \bar{\mathbf{z}}^{\gamma'} \in \mathbb{C}[\mathbf{z}, \bar{\mathbf{z}}]$ is an Hermitian polynomial, i.e., $\bar{g} = g$. The *complex* localizing matrix $\mathbf{M}_r^c(g\mathbf{y})$ associated with g and $\mathbf{y}$ is defined by

$$[\mathbf{M}_r^c(g\,\mathbf{y})]_{\beta,\gamma} := L_{\mathbf{y}}^c(g\,\mathbf{z}^\beta \bar{\mathbf{z}}^\gamma) = \sum_{(\beta',\gamma')} g_{\beta',\gamma'} y_{\beta+\beta',\gamma+\gamma'}, \quad \forall \beta, \gamma \in \mathbb{N}_r^n.$$

Both the complex moment matrix and the complex localizing matrix are Hermitian matrices. Note that a distinguished difference between the real moment matrix and the complex moment matrix is that the former has the Hankel property (i.e., $[\mathbf{M}_r(\mathbf{y})]_{\beta,\gamma}$ is a function of $\beta + \gamma$), whereas the latter does not have.

Let $d_j = \lceil \deg(g_j)/2 \rceil, j \in [m]$ and $r_{\min} = \max\{\lceil \deg(f)/2 \rceil, d_1, \ldots, d_m\}$ as before. Then the complex moment hierarchy indexed by $r \geq r_{\min}$ for CPOP (9.1) is given by

$$\begin{cases} \inf\limits_{\mathbf{y}} & L_{\mathbf{y}}^c(f) \\ \text{s.t.} & \mathbf{M}_r^c(\mathbf{y}) \succeq 0 \\ & \mathbf{M}_{r-d_j}^c(g_j\mathbf{y}) \succeq 0, \quad j \in [m] \\ & y_{0,0} = 1. \end{cases} \tag{9.2}$$

The dual of (9.2) can be formulized as the following HSOS relaxation:

$$
\begin{cases}
\sup_{\sigma_j, b} & b \\
\text{s.t.} & f - b = \sigma_0 + \sigma_1 g_1 + \cdots + \sigma_m g_m \\
& \sigma_j \text{ is an HSOS,} \quad j \in \{0\} \cup [m] \\
& \deg(\sigma_0) \leq 2r, \deg(\sigma_j g_j) \leq 2r, \quad j \in [m].
\end{cases}
\tag{9.3}
$$

Remark 9.1 *In (9.2), the expression "$X \succeq 0$" means an Hermitian matrix X to be PSD. Since popular SDP solvers deal with only real SDPs, it is then necessary to convert this constraint to a constraint involving only real matrices. This can be done by introducing the real part A and the imaginary part B of X respectively such that $X = A + B\mathbf{i}$. Then,*

$$
X \succeq 0 \quad \Longleftrightarrow \quad \begin{bmatrix} A & -B \\ B & A \end{bmatrix} \succeq 0.
$$

Remark 9.2 *The first-order moment-(H)SOS relaxation for QCQP is also known as Shor's relaxation. It was proved in [JM15] that the real Shor's relaxation and the complex Shor's relaxation for homogeneous QCQPs yield the same bound. However, in general the complex hierarchy is weaker (i.e., giving looser bounds) than the real hierarchy at the same relaxation order $r > 1$ as Hermitian sums of squares are a special case of real sums of squares; see [JM18].*

Remark 9.3 *By the complex Positivstellensatz theorem due to D'Angelo and Putinar [DP09], global convergence of the complex hierarchy is guaranteed when a sphere constraint is present.*

As for the usual moment-SOS hierarchy, we can reduce the size of SDP relaxations arising from the complex moment-HSOS hierarchy by exploiting CS and/or TS. The procedures are quite similar. The only significant difference is on the definitions of tsp graphs: in the real case, we use $\mathscr{A} \cup 2\mathbb{N}_r^n$ while in the complex case we use $\mathscr{A}$ instead due to the absence of the Hankel structure of complex moment matrices; see [WM22].

9.2 Applications to optimal power flow

The AC-OPF problem aims to minimize the generation cost of an alternating current transmission network under the physical constraints (Kirchhoff's laws, Ohm's law) as well as operational constraints, which can be formulated as the following POP in complex variables:

$$
\begin{cases}
\displaystyle \inf_{V_i, S_k^g \in \mathbb{C}} & \sum_{k \in G} (\mathbf{c}_{2k}(\Re(S_k^g))^2 + \mathbf{c}_{1k}\Re(S_k^g) + \mathbf{c}_{0k}) \\[2mm]
\text{s.t.} & \angle V_{\text{ref}} = 0 \\[1mm]
& \mathbf{S}_k^{gl} \leq S_k^g \leq \mathbf{S}_k^{gu}, \quad \forall k \in G \\[1mm]
& \mathbf{v}_i^l \leq |V_i| \leq \mathbf{v}_i^u, \quad \forall i \in N \\[1mm]
& \sum_{k \in G_i} S_k^g - \mathbf{S}_i^d - \bar{\mathbf{Y}}_i^s |V_i|^2 = \sum_{(i,j) \in E_i \cup E_i^R} S_{ij}, \quad \forall i \in N \\[1mm]
& S_{ij} = (\bar{\mathbf{Y}}_{ij} - \mathbf{i}\frac{\mathbf{b}_{ij}^c}{2})\frac{|V_i|^2}{|\mathbf{T}_{ij}|^2} - \bar{\mathbf{Y}}_{ij}\frac{V_i \bar{V}_j}{\mathbf{T}_{ij}}, \quad \forall (i,j) \in E \\[1mm]
& S_{ji} = (\bar{\mathbf{Y}}_{ij} - \mathbf{i}\frac{\mathbf{b}_{ij}^c}{2})|V_j|^2 - \bar{\mathbf{Y}}_{ij}\frac{\bar{V}_i V_j}{\bar{\mathbf{T}}_{ij}}, \quad \forall (i,j) \in E \\[1mm]
& |S_{ij}| \leq \mathbf{s}_{ij}^u, \quad \forall (i,j) \in E \cup E^R \\[1mm]
& \boldsymbol{\theta}_{ij}^{\Delta l} \leq \angle(V_i \bar{V}_j) \leq \boldsymbol{\theta}_{ij}^{\Delta u}, \quad \forall (i,j) \in E.
\end{cases}
\tag{9.4}
$$

In (9.4), V_i denotes the voltage, S_k^g denotes the power generation, N denotes the set of buses, and G denotes the set of generators. Besides, $\Re(\cdot)$, $\angle(\cdot)$, $|\cdot|$ stand for the real part, the angle, the magnitude of a complex number, respectively. All symbols in boldface are constants. For a full description on the AC-OPF problem, the reader is referred to [BBC$^+$19]. Note that by introducing real variables for both real and imaginary parts of each complex variable, we can convert the AC-OPF problem to a POP involving only real variables.[1]

To tackle an AC-OPF instance, we first compute a locally optimal solution with nonlinear programming tools whose global optimality is however unknown. And we then rely on certain convex relaxation of (9.4) to certify global optimality of the local solution. Suppose that the optimum reported by the local solver is AC and the optimum of the convex relaxation is opt. The *optimality gap* between the locally optimal solution and the convex relaxation is defined by

$$
\text{gap} := \frac{\text{AC} - \text{opt}}{\text{AC}} \times 100\%.
$$

If the optimality gap is less than 1.00%, then we accept the locally optimal solution as globally optimal.

We perform two classes of numerical experiments on AC-OPF instances issued from PGLiB. For the first class, we compare the complex hierarchy with the real hierarchy in terms of strength and efficiency. The results are reported in Table 9.1 where "mb" denotes the maximal size of PSD blocks involved in the relaxations, "opt" denotes the optimum, "time" denotes

---

[1]The expressions involving angles of complex variables can be converted to polynomials by using $\tan(\angle z) = y/x$ for $z = x + \mathbf{i}y \in \mathbb{C}$.

Table 9.1: The complex versus real hierarchy on AC-OPF instances under typical operating conditions.

case name	order	Complex				Real			
		mb	opt	time	gap	mb	opt	time	gap
30_ieee	1st	8	7.5472e3	0.12	8.06%	8	7.5472e3	0.15	8.06%
	1.5th	12	8.2073e3	0.66	0.02%	22	8.2085e3	0.97	0.00%
39_epri	1st	8	1.3565e4	0.17	2.00%	8	1.3565e4	0.22	2.00%
	1.5th	14	1.3765e4	1.08	0.55%	25	1.3842e4	1.12	0.00%
89_pegase	1st	24	1.0670e5	0.72	0.55%	24	1.0670e5	0.74	0.55%
	1.5th	96	1.0709e5	263	0.19%	184	1.0715e5	1232	0.13%
118_ieee	1st	10	9.6900e4	0.49	0.32%	10	9.6901e4	0.57	0.32%
	1.5th	20	9.7199e4	5.22	0.02%	37	9.7214e4	8.78	0.00%
162_ieee_dtc	1st	28	1.0164e5	1.49	5.96%	28	1.0164e5	1.51	5.96%
	1.5th	40	1.0249e5	17.1	5.17%	74	1.0645e5	87.5	1.51%
179_goc	1st	10	7.5016e5	0.72	0.55%	10	7.5016e5	0.77	0.55%
	1.5th	20	7.5078e5	6.77	0.46%	37	7.5382e5	10.6	0.06%
300_ieee	1st	14	5.5424e5	1.41	1.94%	16	5.5424e5	1.49	1.94%
	1.5th	22	5.6455e5	19.1	0.12%	40	5.6522e5	27.3	0.00%
1354_pegase	1st	26	1.2172e6	10.9	3.30%	26	1.2172e6	13.1	3.30%
	1.5th	26	1.2304e6	255	2.29%	49	1.2514e6	392	0.59%
2869_pegase	1st	26	2.4387e6	47.2	0.98%	26	2.4388e6	67.3	0.97%
	1.5th	98	2.4586e6	1666	0.17%	191	-	-	-

running time in seconds, and "-" indicates an out of memory error. We refer to Shor's relaxation (which applies when we convert (9.4) to a QCQP) as the 1st order relaxation and refer to the minimal initial relaxation defined in Section 8.4 as the 1.5th order relaxation.

As one can see from Table 9.1, the complex 1st order relaxation and the real 1st order relaxation give the same lower bound (up to a given precision) while the former runs slightly faster. The complex 1.5th order relaxation typically gives a looser bound than the real 1.5th order relaxation whereas it runs faster by a factor of $1 \sim 8$. In addition, the 1st order relaxation is able to certify global optimality for 4 out of all 9 instances. For the remaining 5 instances, the complex 1.5th order relaxation is able to certify global optimality for 3 instances and the real 1.5th order relaxation is able to certify global optimality for 4 instances.

As the 1st order relaxation is already able to certify global optimality for a large number of AC-OPF instances, we now focus on more challenging AC-OPF instances for which the 1st order relaxation yields an optimality gap greater than 1.00%. The related data of these selected AC-OPF instances are displayed in Table 9.2, in which "var" denotes the number of variables, "cons" denotes the number of constraints, and "mc" denotes the maximal size of variable cliques.

Table 9.2: The data of selected AC-OPF instances.

case name	var	cons	mc	AC	Shor	
					opt	gap
3_lmbd_api	12	28	6	1.1242e4	1.0417e4	7.34%
5_pjm	20	55	6	1.7552e4	1.6634e4	5.22%
24_ieee_rts_sad	114	315	14	7.6943e4	7.3592e4	4.36%
30_as_api	72	297	8	4.9962e3	4.9256e3	1.41%
73_ieee_rts_sad	344	971	16	2.2775e5	2.2148e5	2.75%
162_ieee_dtc_api	348	1809	21	1.2100e5	1.1928e5	1.42%
240_pserc	766	3322	16	3.3297e6	3.2818e6	1.44%
500_tamu_api	1112	4613	20	4.2776e4	4.2286e4	1.14%
793_goc	1780	7019	18	2.6020e5	2.5636e5	1.47%
1888_rte	4356	18257	26	1.4025e6	1.3748e6	1.97%
3022_goc	6698	29283	50	6.0143e5	5.9278e5	1.44%

Since the real relaxations typically yield tighter lower bounds when the relaxation order is greater than one, we only investigate the real relaxations here. Particularly, we solve the CSSOS hierarchy with $r = 2$ and the CS-TSSOS hierarchy with $r = 2, s = 1$ for these AC-OPF instances, and report the results in Table 9.3. As the table shows, CS-TSSOS is more efficient and scales much better with the problem size than CSSOS. In particular, CS-TSSOS succeeds in reducing the optimality gap to less than 1.00% for all instances.

9.3   Notes and sources

The complex moment-HSOS hierarchy was initially introduced and studied in [JM15, JM18], which has been shown to have advantages over the real hierarchy for certain CPOPs (e.g., a simplified version of the AC-OPF problem).

The AC-OPF is a fundamental problem in power systems, which has been extensively studied in recent years; for a detailed introduction and recent developments, the reader is referred to the survey [BEGL20] and references therein. Since 2006, several convex relaxation schemes (e.g., second-order cone relaxations (SOCR) [Jab06], quadratic convex relaxations (QCR) [CHVH15], tight-and-cheap conic relaxations (TCR) [BALD18] and semidefinite relaxations (SDR) [BWFW08]) have been proposed to provide lower bounds for the AC-OPF which can be then used to certify global optimality of local optimal solutions. While these relaxations (SOCR, QCR,

Table 9.3: The CSSOS versus CS-TSSOS hierarchy on AC-OPF instances.

case name	CSSOS				CS-TSSOS			
	mb	opt	time	gap	mb	opt	time	gap
3_lmbd_api	28	1.1242e4	0.21	0.00%	22	1.1242e4	0.09	0.00%
5_pjm	28	1.7543e4	0.56	0.05%	22	1.7543e4	0.30	0.05%
24_ieee_rts_sad	120	7.6943e4	94.9	0.00%	39	7.6942e4	14.8	0.00%
30_as_api	45	4.9927e3	4.43	0.07%	22	4.9920e3	2.69	0.08%
73_ieee_rts_sad	153	2.2775e5	504	0.00%	44	2.2766e5	71.5	0.04%
162_ieee_dtc_api	253	-	-	-	34	1.2096e5	201	0.03%
240_pserc	153	3.3072e6	585	0.68%	44	3.3042e6	33.9	0.77%
500_tamu_api	231	4.2413e4	3114	0.85%	39	4.2408e4	46.6	0.86%
793_goc	190	2.5938e5	563	0.31%	33	2.5932e5	66.1	0.34%
1888_rte	378	-	-	-	27	1.3953e6	934	0.51%
3022_goc	1326	-	-	-	76	5.9858e5	1886	0.47%

TCR, SDR) could be scalable to problems of large size and prove to be tight for quite a few cases [BBC$^+$19, CHVH15, EDA19], they yield significant optimality gaps on a large number of cases.[2] To tackle these more challenging cases, it is then mandatory to go to higher steps of the moment-SOS hierarchy which can provide tighter lower bounds. Along with this line recently in [GHW$^+$20], Gopinath *et al.* certified 1% global optimality for all AC-OPF instances with up to 300 buses from the AC-OPF library PGLiB using an SDP-based bound tightening approach. Relying on the complex moment-SOS hierarchy combined with a multi-order technique, Josz and Molzahn certified 0.05% global optimality for certain 2000-bus cases on a simplified AC-OPF model [JM18]. A comprehensive numerical study on AC-OPF instances from PGLiB with up to tens of thousands of variables and constraints via the CS-TSSOS hierarchy could be found in [WML22].

---

[2]The reader may find related results on benchmarking SOCR and QR at https://github.com/power-grid-lib/pglib-opf/blob/master/BASELINE.md.

Bibliography

[BALD18]   Christian Bingane, Miguel F Anjos, and Sébastien Le Diga-
            bel. Tight-and-cheap conic relaxation for the ac optimal power
            flow problem. *IEEE Transactions on Power Systems*, 33(6):7131–
            7188, 2018.

[BBC+19]   Sogol Babaeinejadsarookolaee, Adam Birchfield, Richard D
            Christie, Carleton Coffrin, Christopher DeMarco, Ruisheng
            Diao, Michael Ferris, Stephane Fliscounakis, Scott Greene,
            Renke Huang, et al. The power grid library for bench-
            marking AC optimal power flow algorithms. *arXiv preprint
            arXiv:1908.02788*, 2019.

[BEGL20]   Dan Bienstock, Mauro Escobar, Claudio Gentile, and Leo Lib-
            erti. Mathematical programming formulations for the alter-
            nating current optimal power flow problem. *4OR*, 18(3):249–
            292, 2020.

[BWFW08]   Xiaoqing Bai, Hua Wei, Katsuki Fujisawa, and Yong Wang.
            Semidefinite programming for optimal power flow problems.
            International Journal of Electrical Power & Energy Systems, 30(6-
            7):383–392, 2008.

[CHVH15]   Carleton Coffrin, Hassan L Hijazi, and Pascal Van Henten-
            ryck. The QC relaxation: A theoretical and computational
            study on optimal power flow. *IEEE Transactions on Power Sys-
            tems*, 31(4):3008–3018, 2015.

[DP09]   John P D'Angelo and Mihai Putinar. Polynomial optimization
            on odd-dimensional spheres. In *Emerging applications of alge-
            braic geometry*, pages 1–15. Springer, 2009.

[EDA19]   Anders Eltved, Joachim Dahl, and Martin S Andersen. On the
            robustness and scalability of semidefinite relaxation for opti-
            mal power flow problems. *Optimization and Engineering*, pages
            1–18, 2019.

[GHW⁺20] S Gopinath, Hassan L Hijazi, Tillmann Weisser, Harsha Na-garajan, Mertcan Yetkin, Kaarthik Sundar, and Russel W Bent. Proving global optimality of ACOPF solutions. *Electric Power Systems Research*, 189, 2020.

[GMM15] Bissan Ghaddar, Jakub Marecek, and Martin Mevissen. Optimal power flow as a polynomial optimization problem. *IEEE Transactions on Power Systems*, 31(1):539–546, 2015.

[Jab06] Rabih A Jabr. Radial distribution load flow using conic programming. *IEEE transactions on power systems*, 21(3):1458–1459, 2006.

[JM15] Cédric Josz and Daniel K Molzahn. Moment/sum-of-squares hierarchy for complex polynomial optimization. *arXiv preprint arXiv:1508.02068*, 2015.

[JM18] Cédric Josz and Daniel K Molzahn. Lasserre hierarchy for large scale polynomial optimization in real and complex variables. *SIAM Journal on Optimization*, 28(2):1017–1048, 2018.

[WM22] Jie Wang and Victor Magron. Exploiting sparsity in complex polynomial optimization. *Journal of Optimization Theory and Applications*, 192(1):335–359, 2022.

[WML22] Jie Wang, Victor Magron, and Jean B. Lasserre. Certifying global optimality of ac-opf solutions via sparse polynomial optimization. *Electric Power Systems Research*, 213:108683, 2022.

Chapter 10

Exploiting term sparsity in noncommutative polynomial optimization

In this chapter, we generalize the methodology of exploiting TS to noncommutative polynomial optimization. For the sake of conciseness, we consider the problem of eigenvalue optimization over noncommutative polynomials and omit the proofs.

10.1 Eigenvalue optimization with term sparsity

Recall that the eigenvalue optimization problem is defined by

$$\lambda_{\min}(f, \mathfrak{g}) := \inf\{\langle f(\underline{A})\mathbf{v}, \mathbf{v}\rangle : \underline{A} \in \mathcal{D}_{\mathfrak{g}}^{\infty}, \|\mathbf{v}\| = 1\}, \tag{10.1}$$

for $f \in \operatorname{Sym} \mathbb{R}\langle \underline{x}\rangle$ and $\mathfrak{g} = \{g_1, \ldots, g_m\} \subseteq \operatorname{Sym} \mathbb{R}\langle \underline{x}\rangle$. Let

$$\mathscr{A} = \operatorname{supp}(f) \cup \bigcup_{j=1}^{m} \operatorname{supp}(g_j). \tag{10.2}$$

As before, we set $g_0 := 1$, and let $d_j = \lceil \deg(g_j)/2 \rceil$ for $j \in \{0\} \cup [m]$ and $r_{\min} = \max\{\lceil \deg(f)/2 \rceil, d_1, \ldots, d_m\}$. Fixing a relaxation order $r \geq r_{\min}$, we define a graph G_r^{tsp} with nodes $\mathbf{W}_r{}^1$ and edges

$$E(G_r^{\text{tsp}}) = \{\{u, v\} \mid (u, v) \in \mathbf{W}_r \times \mathbf{W}_r, u \neq v, u^\star v \in \mathscr{A} \cup \mathbf{W}_r^2\}, \tag{10.3}$$

---

[1] If $\mathfrak{g} = \varnothing$, then we may replace the monomial basis $\mathbf{W}_r$ with the one returned by the Newton chip method; see [BKP16, §2.3].

where $\mathbf{W}_r^2 := \{u^\star u \mid u \in \mathbf{W}_r\}$. We call G_r^{tsp} the tsp graph associated with the support $\mathscr{A}$.

For a graph $G(V, E)$ with $V \subseteq \langle \underline{x} \rangle$ and $g \in \mathbb{R}\langle \underline{x} \rangle$, let us define

$$\text{supp}_g(G) := \{u^\star wv \mid u = v \in V \text{ or } \{u, v\} \in E, w \in \text{supp}(g)\}. \quad (10.4)$$

Let $G_{r,0}^{(0)} = G_r^{\text{tsp}}$ and $G_{r,j}^{(0)}$ be the empty graph with nodes $V_{r,j} := \mathbf{W}_{r-d_j}$ for $j \in [m]$. Then for each $j \in \{0\} \cup [m]$, we iteratively define a sequence of graphs $(G_{r,j}^{(s)}(V_{r,j}, E_{r,j}^{(s)}))_{s \geq 1}$ via two successive operations:

(1) **support extension**. Let $F_{r,j}^{(s)}$ be the graph with nodes $V_{r,j}$ and

$$E(F_{r,j}^{(s)}) = \{\{u, v\} \mid (u, v) \in V_{r,j} \times V_{r,j}, u \neq v,$$
$$u^\star \text{supp}(g_j)v \cap \bigcup_{j=0}^{m} \text{supp}_{g_j}(G_{r,j}^{(s-1)}) \neq \varnothing\}, \quad (10.5)$$

where $u^\star \text{supp}(g_j)v := \{u^\star wv \mid w \in \text{supp}(g_j)\}$.
(2) **chordal extension**. Let

$$G_{r,j}^{(s)} := (F_{r,j}^{(s)})'. \quad (10.6)$$

By construction, one has $G_{r,j}^{(s)} \subseteq G_{r,j}^{(s+1)}$ for all j, s. Therefore, for every j, the sequence of graphs $(G_{r,j}^{(s)})_{s \geq 1}$ stabilizes after a finite number of steps.

Let $t_j = |\mathbf{W}_{r-d_j}|$ for $j \in \{0\} \cup [m]$. Then by replacing the PSD constraint $\mathbf{M}_{r-d_j}(g_j \mathbf{y}) \succeq 0$ with the weaker constraint $\mathbf{B}_{G_{r,j}^{(s)}} \circ \mathbf{M}_{r-d_j}(g_j \mathbf{y}) \in \Pi_{G_{r,j}^{(s)}}(\mathbf{S}_{t_j}^+)$ for $j \in \{0\} \cup [m]$ in (6.23), we obtain the following series of sparse moment relaxations for (10.1) indexed by $s \geq 1$:

$$\lambda_{\text{ts}}^{r,s}(f, \mathfrak{g}) := \inf_{\mathbf{y}} \quad L_{\mathbf{y}}(f)$$
$$\text{s.t.} \quad \mathbf{B}_{G_{r,0}^{(s)}} \circ \mathbf{M}_r(\mathbf{y}) \in \Pi_{G_{r,0}^{(s)}}(\mathbf{S}_{t_0}^+)$$
$$\mathbf{B}_{G_{r,j}^{(s)}} \circ \mathbf{M}_{r-d_j}(g_j \mathbf{y}) \in \Pi_{G_{r,j}^{(s)}}(\mathbf{S}_{t_j}^+), \quad j \in [m] \quad (10.7)$$
$$y_1 = 1.$$

We call s the *sparse order*. For each $s \geq 1$, the dual of (10.7) reads as

$$\begin{cases} \sup_{G_j, b} \quad b \\ \text{s.t.} \quad \sum_{j=0}^{m} \langle G_j, \mathbf{D}_w^j \rangle + b\delta_{1w} = f_w, \quad \forall w \in \bigcup_{j=0}^{m} \text{supp}_{g_j}(G_{r,j}^{(s)}) \quad (10.8) \\ \quad\quad G_j \in \mathbf{S}_{t_j}^+ \cap \mathbf{S}_{G_{r,j}^{(s)}}, \quad j \in \{0\} \cup [m] \end{cases}$$

where $\{\mathbf{D}_w^j\}_{j,w}$ are appropriate matrices satisfying $\mathbf{M}_{r-d_j}(g_j\mathbf{y}) = \sum_w \mathbf{D}_w^j y_w$. We then call the TS-adapted moment-SOHS relaxations (10.7)–(10.8) the *NCTSSOS hierarchy* associated with (10.1).

Theorem 10.1 *Let* $\{f\} \cup \mathfrak{g} \subseteq \mathrm{Sym}\ \mathbb{R}\langle \underline{x} \rangle$. *Then the following hold:*

(1) *Suppose that* $\mathcal{D}_\mathfrak{g}$ *contains an nc ε-neighborhood of* 0. *Then for all* r, s, *there is no duality gap between* (10.7) *and its dual* (10.8).

(2) *Fixing a relaxation order* $r \geq r_{\min}$, *the sequence* $(\lambda_{ts}^{r,s}(f, \mathfrak{g}))_{s \geq 1}$ *is monotonically nondecreasing and* $\lambda_{ts}^{r,s}(f, \mathfrak{g}) \leq \lambda^r(f, \mathfrak{g})$ *for all s (with* $\lambda^r(f, \mathfrak{g})$ *being defined in* (6.23)*).*

(3) *Fixing a sparse order* $s \geq 1$, *the sequence* $(\lambda_{ts}^{r,s}(f, \mathfrak{g}))_{r \geq r_{\min}}$ *is monotonically nondecreasing.*

(4) *If the maximal chordal extension is chosen in* (10.6), *then* $(\lambda_{ts}^{r,s}(f, \mathfrak{g}))_{s \geq 1}$ *converges to* $\lambda^r(f, \mathfrak{g})$ *in finitely many steps.*

Following from Theorem 10.1, we have the following two-level hierarchy of lower bounds for the optimum $\lambda_{\min}(f, \mathfrak{g})$ of (10.1):

$$
\begin{array}{ccccccc}
\lambda_{ts}^{r_{\min},1}(f, \mathfrak{g}) & \leq & \lambda_{ts}^{r_{\min},2}(f, \mathfrak{g}) & \leq & \cdots & \leq & \lambda^{r_{\min}}(f, \mathfrak{g}) \\
\wedge| & & \wedge| & & & & \wedge| \\
\lambda_{ts}^{r_{\min}+1,1}(f, \mathfrak{g}) & \leq & \lambda_{ts}^{r_{\min}+1,2}(f, \mathfrak{g}) & \leq & \cdots & \leq & \lambda^{r_{\min}+1}(f, \mathfrak{g}) \\
\wedge| & & \wedge| & & & & \wedge| \\
\vdots & & \vdots & & \vdots & & \vdots \\
\wedge| & & \wedge| & & & & \wedge| \\
\lambda_{ts}^{r,1}(f, \mathfrak{g}) & \leq & \lambda_{ts}^{r,2}(f, \mathfrak{g}) & \leq & \cdots & \leq & \lambda^r(f, \mathfrak{g}) \\
\wedge| & & \wedge| & & & & \wedge| \\
\vdots & & \vdots & & \vdots & & \vdots
\end{array}
\qquad (10.9)
$$

Example 10.2 *Consider* $f = 2 - x^2 + xy^2x - y^2 + xyxy + yxyx + x^3y + yx^3 + xy^3 + y^3x$ *and* $\mathfrak{g} = \{1 - x^2, 1 - y^2\}$. *The graph sequence* $(G_{2,0}^{(s)})_{s \geq 1}$ *for f and* $\mathfrak{g}$ *is given in Figure 10.1. In fact the graph sequence* $(G_{2,j}^{(s)})_{s \geq 1}$ *stabilizes at* $s = 2$ *for* $j = 0, 1, 2$ *(with approximately smallest chordal extensions). Using TSSOS, we obtain that* $\lambda_{ts}^{2,1}(f, \mathfrak{g}) \approx -2.55482$, $\lambda_{ts}^{2,2}(f, \mathfrak{g}) = \lambda^2(f, \mathfrak{g}) \approx -2.05111$.

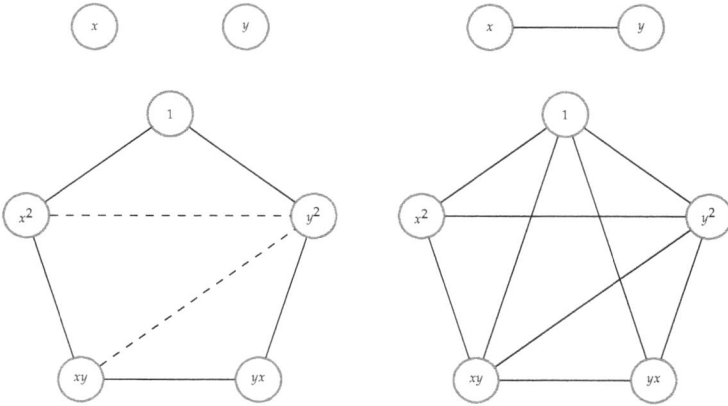

Figure 10.1: The graph sequence $(G_{2,0}^{(s)})_{s\geq 1}$ in Example 10.2: left for $s = 1$; right for $s = 2$. The dashed edges are added after a chordal extension.

10.2   Combining correlative and term sparsity

Combining CS with TS for eigenvalue optimization proceeds in a similar manner as for the commutative case in Section 8.1.

Let $f = \sum_w f_w w \in \mathrm{Sym}\,\mathbb{R}\langle \underline{x}\rangle$ and $\mathfrak{g} = \{g_1,\ldots,g_m\} \subseteq \mathrm{Sym}\,\mathbb{R}\langle \underline{x}\rangle$. Suppose that G^{csp} is the csp graph associated with f and $\mathfrak{g}$, and $(G^{\mathrm{csp}})'$ is a chordal extension of G^{csp}. Let $\{I_k\}_{k\in[p]}$ be the maximal cliques of $(G^{\mathrm{csp}})'$ with cardinality being denoted by $n_k, k \in [p]$. Then the set of variables $\underline{x}$ is decomposed into $\underline{x}(I_1), \underline{x}(I_2),\ldots,\underline{x}(I_p)$. Let $J_1,\ldots,J_p$ be defined as in Section 6.2.

Now we consider the tsp for each subsystem involving the variables $\underline{x}(I_k), k \in [p]$ respectively as follows. Let

$$\mathscr{A} := \mathrm{supp}(f) \cup \bigcup_{j=1}^{m} \mathrm{supp}(g_j) \text{ and } \mathscr{A}_k := \{w \in \mathscr{A} \mid \mathrm{var}(w) \subseteq \underline{x}(I_k)\},$$

$$(10.10)$$

for $k \in [p]$. As before, let $g_0 = 1$, $d_j = \lceil \deg(g_j)/2 \rceil$, $j \in \{0\} \cup [m]$ and $r_{\min} = \max\{\lceil \deg(f)/2 \rceil, d_1,\ldots,d_m\}$. Fix a relaxation order $r \geq r_{\min}$. Let $\mathbf{W}_{r-d_j,k}$ be the standard monomial basis of degree $\leq r - d_j$ with respect to the variables $\underline{x}(I_k)$ and $G_{r,k}^{\mathrm{tsp}}$ be the tsp graph with nodes $\mathbf{W}_{r,k}$ associated with $\mathscr{A}_k$ defined as in Section 10.1. Let $G_{r,k,0}^{(0)} = G_{r,k}^{\mathrm{tsp}}$ and $G_{r,k,j}^{(0)}$ be the empty graph with nodes $V_{r,k,j} := \mathbf{W}_{r-d_j,k}$ for $j \in J_k, k \in [p]$. Letting

$$\mathscr{C}_r^{(s)} := \bigcup_{k=1}^{p} \bigcup_{j\in\{0\}\cup J_k} \mathrm{supp}_{g_j}(G_{r,k,j}^{(s)}),$$

$$(10.11)$$

we iteratively define a sequence of graphs $(G_{r,k,j}^{(s)}(V_{r,k,j}, E_{r,k,j}^{(s)}))_{s\geq 1}$ for each $j \in \{0\} \cup J_k, k \in [p]$ by

$$G_{r,k,j}^{(s)} := (F_{r,k,j}^{(s)})', \tag{10.12}$$

where $F_{r,k,j}^{(s)}$ is the graph with nodes $V_{r,k,j}$ and edges

$$E(F_{r,k,j}^{(s)}) = \{\{u,v\} \mid u \neq v \in V_{r,k,j}, u^{\star} \operatorname{supp}(g_j)v \cap \mathscr{C}_r^{(s-1)} \neq \varnothing\}. \tag{10.13}$$

Let $t_{k,j} = |\mathbf{W}_{r-d_j,k}|$ for all k, j. Then for each $s \geq 1$ (called the *sparse order*), the moment relaxation based on correlative-term sparsity for (10.1) is given by

$$
\begin{aligned}
\lambda_{\text{cs-ts}}^{r,s}(f, \mathfrak{g}) := \quad &\inf_{\mathbf{y}} \quad L_{\mathbf{y}}(f) \\
&\text{s.t.} \quad \mathbf{B}_{G_{r,k,0}^{(s)}} \circ \mathbf{M}_r(\mathbf{y}, I_k) \in \Pi_{G_{r,k,0}^{(s)}}(\mathbf{S}_{r_{k,0}}^+), \quad k \in [p] \\
&\qquad \mathbf{B}_{G_{r,k,j}^{(s)}} \circ \mathbf{M}_{r-d_j}(g_j \mathbf{y}, I_k) \in \Pi_{G_{r,k,j}^{(s)}}(\mathbf{S}_{r_{k,j}}^+), \quad j \in J_k, k \in [p] \\
&\qquad y_1 = 1.
\end{aligned}
\tag{10.14}
$$

For all k, j, let us write $\mathbf{M}_{r-d_j}(g_j \mathbf{y}, I_k) = \sum_w \mathbf{D}_w^{k,j} y_w$ for appropriate matrices $\{\mathbf{D}_w^{k,j}\}_{k,j,w}$. Then for each $s \geq 1$, the dual of (10.14) reads as

$$
\begin{cases}
\sup_{G_{k,j}, b} \quad b \\
\text{s.t.} \quad \sum_{k=1}^p \sum_{j \in \{0\} \cup J_k} \langle G_{k,j}, \mathbf{D}_w^{k,j} \rangle + b\delta_{1w} = f_w, \quad \forall w \in \mathscr{C}_r^{(s)} \\
\qquad G_{k,j} \in \mathbf{S}_{t_{k,j}}^+ \cap \mathbf{S}_{G_{r,k,j}^{(s)}}, \quad j \in \{0\} \cup J_k, k \in [p]
\end{cases}
\tag{10.15}
$$

where $\mathscr{C}_r^{(s)}$ is defined in (10.11).

The properties of the relaxations (10.14)–(10.15) are summarized in the following theorem.

Theorem 10.3 *Assume that $\{f\} \cup \mathfrak{g} \subseteq \operatorname{Sym} \mathbb{R}\langle \underline{x} \rangle$. Then the following hold:*

(1) *Fixing a relaxation order $r \geq r_{\min}$, the sequence $(\lambda_{\text{cs-ts}}^{r,s}(f, \mathfrak{g}))_{s \geq 1}$ is monotonically nondecreasing and $\lambda_{\text{cs-ts}}^{r,s}(f, \mathfrak{g}) \leq \lambda_{\text{cs}}^r(f, \mathfrak{g})$ for all $s \geq 1$ (with $\lambda_{\text{cs}}^r(f, \mathfrak{g})$ being defined in Section 6.25).*

(2) *Fixing a sparse order $s \geq 1$, the sequence $(\lambda_{\text{cs-ts}}^{r,s}(f, \mathfrak{g}))_{r \geq r_{\min}}$ is monotonically nondecreasing.*

(3) *If the maximal chordal extension is chosen in* (10.12), *then* $(\lambda_{cs\text{-}ts}^{r,s}(f,\mathfrak{g}))_{s\geq 1}$ *converges to* $\lambda_{cs}^{r}(f,\mathfrak{g})$ *in finitely many steps.*

10.3 Numerical experiments

In this section, we present numerical results of the proposed NCTSSOS hierarchies. The tool NCTSSOS to implement these hierarchies is available at

https://github.com/wangjie212/NCTSSOS

NCTSSOS employs Mosek as an SDP solver. All numerical examples were computed on an Intel Core i5-8265U@1.60GHz CPU with 8GB RAM memory. In the following, "mb" denotes the maximal size of PSD blocks, "opt" denotes the optimum, "time" denotes running time in seconds, and "-" indicates an out of memory error.

Let **D** be the semialgebraic set defined by $\mathfrak{g} = \{1 - x_1^2, \ldots, 1 - x_n^2, x_1 - 1/3, \ldots, x_n - 1/3\}$, and consider the optimization problem of minimizing the eigenvalue of the nc Broyden banded function on **D**, where the nc Broyden banded function is defined by

$$f_{\text{Bb}}(\mathbf{x}) = \sum_{i=1}^{n} \left(2x_i + 5x_i^3 + 1 - \sum_{j \in J_i} \left(x_j + x_j^2 \right) \right)^{\star}$$
$$\left(2x_i + 5x_i^3 + 1 - \sum_{j \in J_i} \left(x_j + x_j^2 \right) \right),$$

where $J_i = \{j \mid j \neq i, \max(1, i - 5) \leq j \leq \min(n, i + 1)\}$.

We compute $\lambda_{cs\text{-}ts}^{3,1}(f,\mathfrak{g})$ using approximately smallest chordal extensions for TS and present the results in Table 10.1, indicated by "CS+TS". To show the benefits of NCTSSOS against the CS-based approach developed in Chapter 6, we also display the results for the latter approach in the table, indicated by "CS". It is evident from the table that NCTSSOS is much more scalable than the CS-based approach. Actually, the latter can never be executed due to the memory limit even when the problem has only 6 variables.

Now we construct randomly generated examples whose csp graph consists of p maximal cliques of size 15 as follows: let $f = \sum_{k=1}^{p}(h_k + h_k^{\star})/2$ where $h_k \in \mathbb{R}\langle x_{10k-9}, \ldots, x_{10k+5}\rangle$ is a random quartic nc polynomials with 15 terms and coefficients being taken from $[-1, 1]$, and let $\mathfrak{g} = \{g_k\}_{k=1}^{p}$ where $g_k = 1 - x_{10k-9}^2 - \cdots - x_{10k+5}^2$. We consider the eigenvalue minimization problem for f on the multi-ball defined by $\mathfrak{g}$. Let

Table 10.1: The eigenvalue minimization for the nc Broyden banded function on $\mathbf{D}$ with $r = 3, s = 1$.

n	CS+TS			CS		
	mb	opt	time	mb	opt	time
5	11	3.113	0.50	156	3.113	70.7
10	15	3.011	2.78	400	-	-
20	15	9.658	11.4	400	-	-
40	15	22.93	38.1	400	-	-
60	15	36.21	80.5	400	-	-
80	15	49.49	138	400	-	-
100	15	62.77	180	400	-	-

$p = 100, 200, 300, 400$ so that we obtain 4 such instances.[2] We compute the sequence $(\lambda^{r,s}_{\text{cs-ts}}(f, \mathfrak{g}))_{s \geq 1}$ with $r = 2$ and present the results of the first three steps (where we use the maximal chordal extension for the first step and use approximate smallest chordal extensions for the second and third steps, respectively) in Table 10.2. Again we see that NCTSSOS is much more scalable than the CS-based approach.

10.4 Polynomial Bell inequalities

The above framework can be extended to minimize the (normalized) trace of either noncommutative polynomials or a so-called *trace polynomial*. Let us denote by tr the normalized trace operator. A trace polynomial is a polynomial in symmetric noncommutative variables $x_1, \ldots, x_n$ and traces of their products. Thus naturally each trace polynomial f has an adjoint f^*. A *pure trace polynomial* is a trace polynomial that is made only of traces, i.e., has no free variables x_j. For instance, the trace of a trace polynomial is a pure trace polynomial, e.g.,

$$f = x_1 x_2 x_1^2 - \text{tr}(x_2) \, \text{tr}(x_1 x_2) \, \text{tr}(x_1^2 x_2) x_2 x_1,$$
$$\text{tr}(f) = \text{tr}(x_1^3 x_2) - \text{tr}(x_2) \, \text{tr}(x_1 x_2)^2 \, \text{tr}(x_1^2 x_2),$$
$$f^* = x_1^2 x_2 x_1 - \text{tr}(x_2) \, \text{tr}(x_1 x_2) \, \text{tr}(x_1^2 x_2) x_1 x_2.$$

In this section we connect trace polynomial optimization to violations of nonlinear Bell inequalities, outline a few examples and prove the optimal bound on maximal violation of a covariance Bell inequality considered in the quantum information literature.

---

[2]The polynomials are available at `https://wangjie212.github.io/jiewang/code.html`.

Table 10.2: The eigenvalue minimization for randomly generated examples over multi-balls with $r = 2$.

n	s	CS+TS			CS		
		mb	opt	time	mb	opt	time
1005	1	25	−32.58	9.71			
	2	25	−31.91	24.5	241	-	-
	3	25	−31.71	40.9			
2005	1	25	−63.58	33.7			
	2	25	−62.05	85.8	241	-	-
	3	25	−61.76	149			
3005	1	23	−95.73	74.8			
	2	23	−93.13	212	241	-	-
	3	23	−92.71	396			
4005	1	25	−131.1	122			
	2	25	−127.5	375	241	-	-
	3	25	−126.8	687			

We already introduced classical (linear) Bell inequalities in Section 6.7.2, and saw that $v^*(A_1 \otimes B_1 + A_1 \otimes B_2 + A_2 \otimes B_1 - A_2 \otimes B_2)v$ is at most 2 for all separable states $v \in \mathbb{C}^k \otimes \mathbb{C}^k$ and $A_j, B_j \in \mathbb{C}^{k \times k}$ satisfying $A_j^* = A_j$, $A_j^2 = I$, $B_j^* = B_j$, $B_j^2 = I$. Tsirelson's bound implies that the value is at most $2\sqrt{2}$, which is attained in particular when $k = 2$ and $v = \frac{1}{\sqrt{2}}(e_1 \otimes e_1 + e_2 \otimes e_2)$, with (e_1, e_2) being an orthonormal basis of $\mathbb{C}^2$. In general, if v_k is the generalized Bell state,

$$v_k = \frac{1}{\sqrt{k}} \sum_{j=1}^{k} e_j \otimes e_j \in \mathbb{R}^k \otimes \mathbb{R}^k,$$

which is a maximally entangled bipartite state on $\mathbb{C}^k \otimes \mathbb{C}^k$, unique up to bipartite unitary equivalence, then

$$v_k^*(X \otimes Y)v_k = \text{tr}(XY) \tag{10.16}$$

for all $X, Y \in \mathbb{S}_k$.

While linear Bell inequalities are linear in expectation values of products of observables, polynomial Bell inequalities contain multivariate polynomials in expectation values of products of observables. For this reason, noncommutative polynomial optimization is not suitable for studying violations of nonlinear Bell inequalities. In contrast, trace polynomial optimization gives upper bounds on violations of polynomial Bell inequalities, at least for certain families of states, e.g., the maximally entangled bipartite states via (10.16). Consider a simple quadratic Bell inequality

$$\left(v^*(A_1 \otimes B_2 + A_2 \otimes B_1)v\right)^2 + \left(v^*(A_2 \otimes B_1 - A_2 \otimes B_2)v\right)^2 \leq 4. \quad (10.17)$$

An automatized proof of (10.17) for maximally entangled states of arbitrary dimension can be obtained by solving the optimization problem:

$$\begin{cases} \sup & (\operatorname{tr}(a_1 b_2 + a_2 b_1))^2 + (\operatorname{tr}(a_1 b_1 - a_2 b_2))^2 \\ \text{s.t.} & a_j^2 = 1, b_j^2 = 1 \text{ for } j = 1, 2. \end{cases} \quad (10.18)$$

We compare the value of the dense relaxation of (10.18) with the ones obtained after exploiting TS. At relaxation order $r = 2$, we obtain a bound of 4 (the optimal one) in the dense setting. The number of SDP equality constraints is 222 and the size of the SDP matrix is 53. At the sparse order $s = 1$ and using the maximal chordal extension, we obtain the same bound but the maximal block size is only 6 and the number of equality constraints is only 30.

Another class of polynomial Bell inequalities arises from covariances of quantum correlations. Let $\operatorname{cov}_v(X, Y) := v^*(X \otimes Y)v - v^*(X \otimes I)v \cdot v^*(I \otimes Y)v.$ and let us show that the value of

$$\begin{aligned} &\operatorname{cov}_v(A_1, B_1) + \operatorname{cov}_v(A_1, B_2) + \operatorname{cov}_v(A_1, B_3) \\ &+ \operatorname{cov}_v(A_2, B_1) + \operatorname{cov}_v(A_2, B_2) - \operatorname{cov}_v(A_2, B_3) \\ &+ \operatorname{cov}_v(A_3, B_1) - \operatorname{cov}_v(A_3, B_2) \end{aligned} \quad (10.19)$$

is at most 5 for every maximally entangled state. Let

$$\begin{aligned} t = {} & \operatorname{tr}(a_1 b_1) - \operatorname{tr}(a_1)\operatorname{tr}(b_1) + \operatorname{tr}(a_1 b_2) - \operatorname{tr}(a_1)\operatorname{tr}(b_2) + \operatorname{tr}(a_1 b_3) + \operatorname{tr}(a_2 b_1) \\ & - \operatorname{tr}(a_1)\operatorname{tr}(b_3) - \operatorname{tr}(a_2)\operatorname{tr}(b_1) + \operatorname{tr}(a_2 b_2) - \operatorname{tr}(a_2)\operatorname{tr}(b_2) - \operatorname{tr}(a_2 b_3) \\ & + \operatorname{tr}(a_2)\operatorname{tr}(b_3) + \operatorname{tr}(a_3 b_1) - \operatorname{tr}(a_3)\operatorname{tr}(b_1) - \operatorname{tr}(a_3 b_2) + \operatorname{tr}(a_3)\operatorname{tr}(b_2). \end{aligned}$$

The dense relaxation of

$$\sup t \text{ s.t. } a_j^2 = 1, b_j^2 = 1 \text{ for } j = 1, 2, 3 \quad (10.20)$$

with $r = 2$ returns 5. Therefore the value of (10.19) is at most 5 for every maximally entangled state, regardless of the local dimension k. This dense relaxation involves 1010 SDP equality constraints and an SDP matrix of size 115. Exploiting TS at the sparse order $s = 1$ with the maximal chordal extension yields the same bound but requires to solve an SDP with maximal block size 83 and 797 equality constraints. We can obtain an even better computational gain by relying on approximately smallest chordal extensions as it yields an SDP with maximal block size 25 and only 115 equality constraints, 10 times less compared to the dense case. We refer the interested programmer to the Julia scripts displayed in Appendix B.2, allowing one to retrieve these results.

10.5   Notes and sources

The material from this chapter is mainly issued from [WM21]. Our framework can be extended to minimize the trace of a noncommutative polynomial over a noncommutative semialgebraic set; see [WM21, §5]. Optimization over trace polynomials has been recently developed in [KMV22], where the authors present a novel Positivstellensatz certifying positivity of trace polynomials subject to trace constraints, and a hierarchy of semidefinite relaxations converging monotonically to the optimum of a trace polynomial subject to tracial constraints is provided. Trace polynomial optimization can be used to detect entanglement in multipartite Werner states [HKMV21]. The interested reader can find more information about bilocal models in [BRGP12, Cha16], covariance of quantum correlations in [PHBB17] and detection of partial separability in [Uff02]. Inequality (10.17) is given in [Uff02], where it is shown to hold for all separable states, and for all 2-dimensional states. In [NKI02], (10.17) is shown to hold for arbitrary states, meaning it admits no quantum violations.

In [PHBB17] it is shown that while the value of (10.19) is at most $\frac{9}{2}$ for separable states, it attains the value 5 with the Bell state v_2. The authors also performed extensive numerical search within entangled states for local dimensions $k \leq 5$, but no higher value of (10.19) was found. They leave it as an open question whether higher dimensional entangled states could lead to larger violations [PHBB17, Appendix D.1(b)].

Bibliography

[BKP16]   Sabine Burgdorf, Igor Klep, and Janez Povh. *Optimization of polynomials in non-commuting variables*. SpringerBriefs in Mathematics. Springer, [Cham], 2016.

[BRGP12] Cyril Branciard, Denis Rosset, Nicolas Gisin, and Stefano Pironio. Bilocal versus nonbilocal correlations in entanglement-swapping experiments. *Phys. Rev. A*, 85:032119, Mar 2012.

[Cha16]   Rafael Chaves. Polynomial Bell inequalities. *Phys. Rev. Lett.*, 116(1):010402, 6, 2016.

[HKMV21] Felix Huber, Igor Klep, Victor Magron, and Jurij Volčič. Dimension-free entanglement detection in multipartite werner states. *arXiv preprint arXiv:2108.08720*, 2021.

[KMV22]   Igor Klep, Victor Magron, and Jurij Volčič. Optimization over trace polynomials. In *Annales Henri Poincaré*, volume 23, pages 67–100. Springer, 2022.

[NKI02]   Koji Nagata, Masato Koashi, and Nobuyuki Imoto. Configuration of separability and tests for multipartite entanglement in Bell-type experiments. *Phys. Rev. Lett.*, 89(26):260401, 4, 2002.

[PHBB17] Victor Pozsgay, Flavien Hirsch, Cyril Branciard, and Nicolas Brunner. Covariance Bell inequalities. *Phys. Rev A*, 96(6):062128, 13, 2017.

[Uff02]   Jos Uffink. Quadratic Bell inequalities as tests for multipartite entanglement. *Phys. Rev. Lett.*, 88(23):230406, 4, 2002.

[WM21]   Jie Wang and Victor Magron. Exploiting term sparsity in non-commutative polynomial optimization. *Computational Optimization and Applications*, 80(2):483–521, 2021.

Chapter 11

Application in stability of control systems

The concept of joint spectral radius (JSR) can be viewed as a generalization of the usual spectral radius to the case of multiple matrices. The exact computation and even the approximation of the JSR for a set of matrices are, however, notoriously difficult. In this chapter, we discuss how to efficiently compute upper bounds on JSR via the SOS approach when the matrices possess certain sparsity.

11.1 Approximating JSR via SOS relaxations

The JSR for a set of matrices $\mathcal{A} = \{A_1, \ldots, A_m\} \subseteq \mathbb{R}^{n \times n}$ is given by

$$\rho(\mathcal{A}) := \lim_{k \to \infty} \max_{\sigma \in \{1, \ldots, m\}^k} \|A_{\sigma_1} A_{\sigma_2} \cdots A_{\sigma_k}\|^{\frac{1}{k}}. \tag{11.1}$$

Note that the value of $\rho(\mathcal{A})$ is independent of the choice of the norm used in (11.1). Parrilo and Jadbabaie proposed to compute a sequence of upper bounds for $\rho(\mathcal{A})$ via SOS relaxations. The underlying idea is based on the following theorem.

Theorem 11.1 ([PJ08], Theorem 2.2) *Let $\mathcal{A} = \{A_1, \ldots, A_m\} \subseteq \mathbb{R}^{n \times n}$ be a set of matrices, and let p be a strictly positive form of degree $2r$ that satisfies*

$$p(A_i \mathbf{x}) \leq \gamma^{2r} p(\mathbf{x}), \quad \forall \mathbf{x} \in \mathbb{R}^n, \quad i = 1, \ldots, m.$$

Then, $\rho(\mathcal{A}) \leq \gamma$.

By replacing positive forms with more tractable SOS forms, Theorem 11.1 immediately suggests the following SOS relaxations indexed by

$r \in \mathbb{N}^*$ to compute a sequence of upper bounds for $\rho(\mathcal{A})$:

$$\rho_{SOS,2r}(\mathcal{A}) := \inf_{p \in \mathbb{R}[x]_{2r}, \gamma} \gamma$$
$$\text{s.t.} \quad p(\mathbf{x}) - \|\mathbf{x}\|_2^{2r} \in \Sigma_{n,2r}$$
$$\gamma^{2r} p(\mathbf{x}) - p(A_i \mathbf{x}) \in \Sigma_{n,2r}, \quad i = 1, \dots, m.$$

(11.2)

The term $"\|\mathbf{x}\|_2^{2r}"$ appearing in the first constraint of (11.2) is added to make sure that p is strictly positive. The optimization problem (11.2) can be solved via SDP by bisection on γ. It was shown in [PJ08] that the upper bound $\rho_{SOS,2r}(\mathcal{A})$ satisfies the inequalities stated in the following theorem.

Theorem 11.2 ([PJ08]) *Let $\mathcal{A} = \{A_1, \dots, A_m\} \subseteq \mathbb{R}^{n \times n}$. For any integer $r \geq 1$, one has $m^{-\frac{1}{2r}} \rho_{SOS,2r}(\mathcal{A}) \leq \rho(\mathcal{A}) \leq \rho_{SOS,2r}(\mathcal{A})$.*

It is immediate from Theorem 11.2 that $(\rho_{SOS,2r}(\mathcal{A}))_{r \geq 1}$ converges to $\rho(\mathcal{A})$ as r goes to infinity.

11.2   The SparseJSR Algorithm

In this section, we propose an algorithm SparseJSR for bounding JSR from above based on the sparse SOS decomposition when the matrices $\mathcal{A}$ possess certain sparsity. The first step is to establish a hierarchy of sparse supports for the auxiliary form $p(\mathbf{x})$ used in the SOS program (11.2).

Let us fix a relaxation order r. Let $p_0(\mathbf{x}) = \sum_{i=1}^n c_i x_i^{2r}$ with random coefficients $c_i \in (0,1)$ and let $\mathscr{A}^{(0)} = \text{supp}(p_0)$. Then for $s \in \mathbb{N}^*$, we iteratively define

$$\mathscr{A}^{(s)} := \mathscr{A}^{(s-1)} \cup \bigcup_{i=1}^m \text{supp}(p_{s-1}(A_i \mathbf{x})),$$

(11.3)

where $p_{s-1}(\mathbf{x}) = \sum_{\alpha \in \mathscr{A}^{(s-1)}} c_\alpha \mathbf{x}^\alpha$ with random coefficients $c_\alpha \in (0,1)$. Note that here the particular form of $p_0(\mathbf{x})$ is chosen such that $\mathscr{A}^{(s)}$ contains all possible homogeneous monomials of degree $2r$ that are "compatible" with the couplings between variables $x_1, \dots, x_n$ introduced by the mappings $\mathbf{x} \mapsto A_i \mathbf{x}$ for all i. It is clear that

$$\mathscr{A}^{(1)} \subseteq \cdots \subseteq \mathscr{A}^{(s)} \subseteq \mathscr{A}^{(s+1)} \subseteq \cdots \subseteq \mathbb{N}_{2r}^n$$

(11.4)

and the sequence $(\mathscr{A}^{(s)})_{s \geq 1}$ stabilizes in finitely many steps. We emphasis that it is not guaranteed a hierarchy of sparse supports is always retrieved by (11.4) even if all A_i are sparse. For instance, if some matrix $A_i \in \mathcal{A}$ has a fully dense row, then by definition, one immediately has $\mathscr{A}^{(1)} = \mathbb{N}_{2r}^n$. In this case, the sparsity of $\mathcal{A}$ cannot be exploited by the present method.

On the other hand, if the matrices in $\mathcal{A}$ have common zero columns, then a hierarchy of sparse supports must be retrieved by (11.4) as shown in the next proposition.

Proposition 11.3 Let $\mathcal{A} = \{A_1, \ldots, A_m\} \subseteq \mathbb{R}^{n \times n}$ and assume that the matrices in $\mathcal{A}$ have common zero columns indexed by $J \subseteq [n]$. Let $\tilde{\mathbb{N}}_{2r}^{n-|J|} :=$ $\{(\alpha_i)_{i \in [n]} \in \mathbb{N}^n \mid (\alpha_i)_{i \in [n] \setminus J} \in \mathbb{N}_{2r}^{n-|J|}, \alpha_i = 0 \text{ for } i \in J\}$ and $\mathbf{b}_j := \{(\alpha_i)_{i \in [n]} \in \mathbb{N}^n \mid \alpha_j = 2r, \alpha_i = 0 \text{ for } i \neq j\}$ for $j \in [n]$. Then $\mathscr{A}^{(s)} \subseteq \tilde{\mathbb{N}}_{2r}^{n-|J|} \cup \{\mathbf{b}_j\}_{j \in J}$ for all $s \geq 1$.

PROOF Let us do induction on s. It is obvious that $\mathscr{A}^{(0)} \subseteq \tilde{\mathbb{N}}_{2r}^{n-|J|} \cup \{\mathbf{b}_j\}_{j \in J}$. Now assume $\mathscr{A}^{(s)} \subseteq \tilde{\mathbb{N}}_{2r}^{n-|J|} \cup \{\mathbf{b}_j\}_{j \in J}$ for some $s \geq 0$. Since the variables effectively involved in $p_s(A_j \mathbf{x})$ are contained in $\{x_i\}_{i \in [n] \setminus J}$, we have $\text{supp}(p_s(A_j \mathbf{x})) \subseteq \tilde{\mathbb{N}}_{2r}^{n-|J|}$ for $j = 1, \ldots, m$. This combined with the induction hypothesis yields $\mathscr{A}^{(s+1)} \subseteq \tilde{\mathbb{N}}_{2r}^{n-|J|} \cup \{\mathbf{b}_j\}_{j \in J}$ as desired. $\quad\square$

For each $s \geq 1$, by restricting $p(\mathbf{x})$ to forms with the sparse support $\mathscr{A}^{(s)}$, (11.2) now reads as

$$
\begin{cases}
\inf_{p \in \mathbb{R}[\mathscr{A}^{(s)}], \gamma} & \gamma \\
\text{s.t.} & p(\mathbf{x}) - \|\mathbf{x}\|_2^{2r} \in \Sigma_{n,2r} \\
& \gamma^{2r} p(\mathbf{x}) - p(A_i \mathbf{x}) \in \Sigma_{n,2r}, \quad i \in [m].
\end{cases}
\tag{11.5}
$$

Let $\mathscr{A}_i^{(s)} = \mathscr{A}^{(s)} \cup \text{supp}(p_s(A_i \mathbf{x})$ for $i = 1, \ldots, m$. In order to exploit the sparsity present in (11.5), for a sparse support $\mathscr{A} \subseteq \mathbb{N}_{2r}^n$ we define

$$
\Sigma(\mathscr{A}) := \left\{ f \in \mathbb{R}[\mathscr{A}] \mid \exists \mathbf{G} \in \mathbb{S}_{|\mathscr{B}|}^+ \cap S_{G^{\text{tsp}}} \text{ s.t. } f = (\mathbf{x}^{\mathscr{B}})^{\mathsf{T}} \mathbf{G} \mathbf{x}^{\mathscr{B}} \right\}, \tag{11.6}
$$

where $\mathscr{B}$ is a monomial basis and G^{tsp} is the tsp graph with respect to $\mathscr{B}$ defined as in Section 7.1. Then by replacing $\Sigma_{n,2d}$ with $\Sigma(\mathscr{A}^{(s)})$ or $\Sigma(\mathscr{A}_i^{(s)})$ in (11.5), we therefore obtain a hierarchy of sparse SOS relaxations indexed by s for a fixed r:

$$
\rho_{s,2r}(\mathcal{A}) := \inf_{p \in \mathbb{R}[\mathscr{A}^{(s)}], \gamma} \gamma
$$
$$
\text{s.t.} \quad p(\mathbf{x}) - \|\mathbf{x}\|_2^{2r} \in \Sigma(\mathscr{A}^{(s)})
$$
$$
\gamma^{2r} p(\mathbf{x}) - p(A_i \mathbf{x}) \in \Sigma(\mathscr{A}_i^{(s)}), \quad i \in [m].
\tag{11.7}
$$

As in the dense case, the optimization problem (11.7) can be solved via SDP by bisection on γ. We call the index s the *sparse order* of (11.7), and we get a hierarchy of upper bounds on the JSR indexed by the sparse order s when the relaxation order r is fixed.

Theorem 11.4 *Let* $\mathcal{A} = \{A_1, \ldots, A_m\} \subseteq \mathbb{R}^{n \times n}$. *For any integer* $r \geq 1$, *one has* $\rho_{SOS,2r}(\mathcal{A}) \leq \cdots \leq \rho_{s,2r}(\mathcal{A}) \leq \cdots \leq \rho_{2,2r}(\mathcal{A}) \leq \rho_{1,2r}(\mathcal{A})$.

PROOF For any fixed $r \in \mathbb{N}^*$, because of (11.4), it is clear that the feasible set of (11.7) with the sparse order s is contained in the feasible set of (11.7) with the sparse order $s + 1$, which is in turn contained in the feasible set of (11.2). This yields the desired conclusion.

Therefore, the algorithm SparseJSR computes a nonincreasing sequence of upper bounds for the JSR of a tuple of matrices via solving (11.7) for any fixed r. By tuning the relaxation order r and the sparse order s, SparseJSR offers a trade-off between the computational cost and the quality of the obtained upper bound.

11.3   Numerical experiments

In this section, we present numerical experiments for the proposed algorithm SparseJSR, which was implemented in the Julia package also named `SparseJSR`. In Appendix B.3, we provide a Julia script to illustrate how to use `SparseJSR`.

The examples were computed on an Intel Core i5-8265U@1.60GHz CPU with 8GB RAM memory, where the sparse order s was set to 1, the tolerance for bisection was set to $\varepsilon = 1 \times 10^{-5}$, and the initial interval for bisection was set to $[0, 2]$. To measure the quality of upper bounds (ub) provided by `SparseJSR`, we also compute lower bounds (lb) on JSR using Gripenberg's algorithm [Gri96]. In the following, "mb" denotes the maximal size of PSD blocks, "time" denotes running time in seconds, $*$ indicates running time exceeding one hour, and "-" indicates an out of memory error.

11.3.1   Randomly generated examples

We generate random sparse matrices as follows:[1] generate a random directed graph G with n nodes and $n + 10$ edges; for each edge (i, j) of G, put a random number in $[-1, 1]$ on the position (i, j) of the matrix and put zeros on the other positions. We compute an upper bound on JSR for pairs of such matrices using the first-order SOS relaxation and report the results in Table 11.1. It is evident that the sparse approach is much more efficient than the dense approach. For instance, the dense approach takes

---

[1] Available at https://wangjie212.github.io/jiewang/code.html.

over 3600 s when the size of matrices is greater than 100 while the sparse approach can handle matrices of size 120 within 12 s. The upper bound produced by the sparse approach is slightly weaker, but is still close to the lower bound.

Table 11.1: Randomly generated examples with $r = 1$ and $m = 2$.

n	lb	Sparse			Dense		
		time	ub	mb	time	ub	mb
20	0.7894	0.74	0.8192	10	1.88	0.7967	20
40	0.9446	2.68	0.9446	14	25.6	0.9446	40
60	0.7612	3.64	0.7843	13	171	0.7612	60
80	0.9345	5.95	0.9399	15	743	0.9345	80
100	0.8642	8.15	0.9132	13	2568	0.8659	100
120	0.7483	11.7	0.7735	16	*	*	*

11.3.2 Examples from control systems

We consider examples from [MHMJZ20], where the dynamics of closed-loop systems are given by the combination of a plant and a one-step delay controller that stabilizes the plant. The closed-loop system evolves according to either a completed or a missed computation. In the case of a deadline hit, the closed-loop state matrix is A_H. In the case of a deadline miss, the associated closed-loop state matrix is A_M. The computational platform (hardware and software) ensures that no more than $m - 1$ deadlines are missed consecutively. The set of possible realisations $\mathcal{A}$ of such a system contains either a single hit or at most $m - 1$ misses followed by a hit, namely $\mathcal{A} := \{A_H A_M^i \mid 0 \leq i \leq m - 1\}$. Then, the closed-loop system that can switch between the realisations included in $\mathcal{A}$ is asymptotically stable if and only if $\rho(\mathcal{A}) < 1$. This gives an indication for scheduling and control co-design, in which the hardware and software platform must guarantee that the maximum number of deadlines missed consecutively does not interfere with stability requirements.

In Tables 11.2 and 11.3, we report the results obtained for various control systems with n states, under $m - 1$ deadline misses, by applying the dense and sparse SOS approaches with relaxation orders $r = 1$ and $r = 2$, respectively. The examples are randomly generated, i.e., our script generates a random system and then tries to control it.[2]

In Table 11.2, we fix $m = 5$ and vary n from 20 to 110. For these examples, surprisingly the dense and sparse approaches with the relaxation

---

[2]Available at https://wangjie212.github.io/jiewang/code.html.

order $r = 1$ always produce the same upper bounds. As one can see from the table, the sparse approach is more efficient and scalable than the dense one.

In Table 11.3, we vary m from 3 to 11 and vary n from 8 to 24. The column "ub" indicates the upper bound given by the dense approach with the relaxation order $r = 1$. For these examples, with the relaxation order $r = 2$, the sparse approach produces upper bounds that are very close to those given by the dense approach. And again the sparse approach is more efficient and scalable than the dense one.

Table 11.2: Results for control systems with $r = 1$ and $m = 5$.

n	lb	Sparse			Dense		
		time	ub	mb	time	ub	mb
30	1.4682	4.30	1.5132	14	57.8	1.5131	30
30	1.0924	4.42	1.0961	14	65.4	1.0961	30
50	1.3153	17.3	1.3248	18	660	1.3248	50
50	1.1884	17.5	1.1884	18	680	1.1884	50
70	1.8135	54.2	1.8578	22	*	*	*
70	1.2727	53.9	1.2727	22	*	*	*
90	1.8745	133	1.9020	26	*	*	*
90	1.4452	132	1.4452	26	*	*	*
110	2.3597	280	2.3943	30	-	-	-
110	1.5753	287	1.5753	30	-	-	-

Table 11.3: Results for control systems with $r = 2$.

m	n	lb	ub	Sparse			Dense		
				time	ub	mb	time	ub	mb
3	8	0.7218	0.7467	0.60	0.7310	10	13.4	0.7305	36
4	10	0.7458	0.7738	0.75	0.7564	10	107	0.7554	55
5	12	0.8601	0.8937	1.08	0.8706	10	1157	0.8699	78
6	14	0.7875	0.8107	1.32	0.7958	10	*	*	*
7	16	1.1110	1.1531	1.81	1.1182	10	-	-	-
8	18	1.0487	1.0881	2.05	1.0569	10	-	-	-
9	20	0.7570	0.7808	2.52	0.7660	10	-	-	-
10	22	0.9911	1.0315	2.70	1.0002	10	-	-	-
11	24	0.7339	0.7530	3.67	0.7418	10	-	-	-

11.4   Notes and sources

The material from this chapter is issued from [WMM21]. The JSR was first introduced by Rota and Strang in [RS60] and since then has found applications in many areas such as the stability of switched linear dynamical systems, the continuity of wavelet functions, combinatorics and language theory, the capacity of some codes, the trackability of graphs. We refer the reader to [Jun09] for a survey of the theory and applications of JSR.

Bibliography

[Gri96]   Gustaf Gripenberg. Computing the joint spectral radius. *Linear Algebra and its Applications*, 234:43–60, 1996.

[Jun09]   Raphaël Jungers. *The joint spectral radius: theory and applications*, volume 385. Springer Science & Business Media, 2009.

[MHMJZ20] Martina Maggio, Arne Hamann, Eckart Mayer-John, and Dirk Ziegenbein. Control-system stability under consecutive deadline misses constraints. In *32nd Euromicro Conference on Real-Time Systems (ECRTS 2020)*. Schloss Dagstuhl-Leibniz-Zentrum für Informatik, 2020.

[PJ08]   Pablo A Parrilo and Ali Jadbabaie. Approximation of the joint spectral radius using sum of squares. *Linear Algebra and its Applications*, 428(10):2385–2402, 2008.

[RS60]   Gian-Carlo Rota and W Strang. A note on the joint spectral radius. In *Gian-Carlo Rota on Analysis and Probability: Selected Papers and Commentaries*, 1960.

[WMM21]   Jie Wang, Martina Maggio, and Victor Magron. Sparsejsr: A fast algorithm to compute joint spectral radius via sparse sos decompositions. In *2021 American Control Conference (ACC)*, pages 2254–2259. IEEE, 2021.

Chapter 12

Miscellaneous

The goal of this chapter is to propose alternative schemes to methods based on sparse SOS polynomials. First, we focus in Section 12.1 on sum of nonnegative circuits (SONC) polynomials, a set of new nonnegativity certificates, independent of the set of SOS polynomials described earlier in this book. We present a characterization of SONC polynomials in terms of sums of binomial squares with rational exponents. Next, we present in Section 12.2 a framework to speed-up the resolution of SDP relaxations arising from the moment-SOS hierarchy. This framework is based on the use of first-order methods.

12.1 Nonnegative circuits and binomial squares

A lattice point $\alpha \in \mathbb{N}^n$ is said to be *even* if it is in $(2\mathbb{N})^n$. A subset $\mathscr{T} = \{\alpha_1, \ldots, \alpha_m\} \subseteq (2\mathbb{N})^n$ is called a *trellis* when $\mathscr{T}$ comprises the vertices of a simplex. Given a trellis $\mathscr{T}$, a *circuit polynomial* is of the form $\sum_{\alpha \in \mathscr{T}} c_\alpha \mathbf{x}^\alpha - d\mathbf{x}^\beta \in \mathbb{R}[\mathbf{x}]$, where $c_\alpha > 0$ for all $\alpha \in \mathscr{T}$, and β lies in the relative interior of the simplex associated to $\mathscr{T}$. The name "circuit polynomial" stems from the fact that $(\mathscr{T}, \beta)$ consists of a circuit. Given a circuit polynomial $f = \sum_{\alpha \in \mathscr{T}} c_\alpha \mathbf{x}^\alpha - d\mathbf{x}^\beta \in \mathbb{R}[\mathbf{x}]$, there exist unique barycentric coordinates $(\lambda)_{j=1}^m$ satisfying

$$\beta = \sum_{j=1}^m \lambda_j \alpha(j) \text{ with } \lambda_j > 0 \text{ and } \sum_{j=1}^m \lambda_j = 1. \tag{12.1}$$

We then define the related circuit number as $\Theta_f = \prod_{j=1}^m \left(c_{\alpha(j)} / \lambda_j \right)^{\lambda_j}$.

Circuit polynomials are proper building blocks for nonnegativity certificates since the circuit number alone determines whether they are nonnegative.

Theorem 12.1 *A circuit polynomial f is nonnegative if and only if f is a sum of monomial squares or $|d| \leq \Theta_f$.*

Example 12.2 *Let $f = x_1^4 x_2^2 + x_1^2 x_2^4 + 1 - 3x_1^2 x_2^2$ be the Motzkin polynomial and $\mathcal{T} = \{\alpha_1 = (0,0), \alpha_2 = (4,2), \alpha_3 = (2,4)\}$, $\beta = (2,2)$. Then $\beta = \frac{1}{3}\alpha_1 + \frac{1}{3}\alpha_2 + \frac{1}{3}\alpha_3$.*

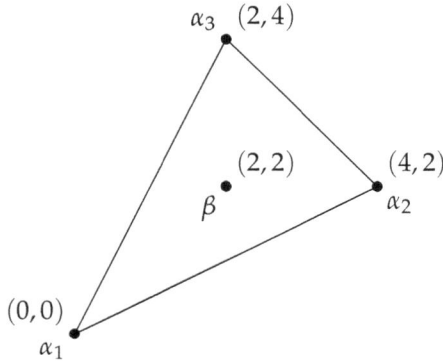

One easily checks that $|-3| \leq \Theta_f = 3$, proving that f is nonnegative.

An explicit representation of a polynomial being a SONC provides a certificate for its nonnegativity, which is called a SONC decomposition. Polynomial optimization using SONC decompositions can be solved by means of geometric programming. Here we introduce a potentially cheaper approach based on second-order cone programming.

For a subset of points $M \subseteq \mathbb{N}^n$, let

$$\overline{A}(M) := \left\{ \frac{1}{2}(\mathbf{v} + \mathbf{w}) \mid \mathbf{v} \neq \mathbf{w}, \mathbf{v}, \mathbf{w} \in M \cap (2\mathbb{N})^n \right\}$$

be the set of averages of distinct even points in M. For a trellis $\mathcal{T}$, we call M a $\mathcal{T}$-*mediated set* if $\mathcal{T} \subseteq M \subseteq \overline{A}(M) \cup \mathcal{T}$.

Theorem 12.3 *Let $f = \sum_{\alpha \in \mathcal{T}} c_\alpha \mathbf{x}^\alpha - d\mathbf{x}^\beta \in \mathbb{R}[\mathbf{x}]$ with $d \neq 0$ be a nonnegative circuit polynomial. Then f is a sum of binomial squares if and only if there exists a $\mathcal{T}$-mediated set containing β. Moreover, suppose that β belongs to a $\mathcal{T}$-mediated set M and for each $\mathbf{u} \in M \setminus \mathcal{T}$, let us write $\mathbf{u} = \frac{1}{2}(\mathbf{v_u} + \mathbf{w_u})$ for some $\mathbf{v_u} \neq \mathbf{w_u} \in M \cap (2\mathbb{N})^n$. Then we can rewrite f as $f = \sum_{\mathbf{u} \in M \setminus \mathcal{T}} (a_\mathbf{u} \mathbf{x}^{\frac{1}{2}\mathbf{v_u}} - b_\mathbf{u} \mathbf{x}^{\frac{1}{2}\mathbf{w_u}})^2$ for some $a_\mathbf{u}, b_\mathbf{u} \in \mathbb{R}$.*

By Theorem 12.3, to represent a nonnegative circuit polynomial as a sum of binomial squares, we need to first decide whether there exists a $\mathscr{T}$-mediated set containing a given lattice point, and then compute one if there exists. However, such a $\mathscr{T}$-mediated set may not exist in general. In order to circumvent this obstacle, we introduce the concept of $\mathscr{T}$-rational mediated sets as a replacement of $\mathscr{T}$-mediated sets by admitting rational numbers in coordinates.

Concretely, for a subset of points $M \subseteq \mathbb{Q}^n$, let us define

$$\widetilde{A}(M) := \left\{ \frac{1}{2}(\mathbf{v} + \mathbf{w}) \mid \mathbf{v} \neq \mathbf{w}, \mathbf{v}, \mathbf{w} \in M \right\}$$

as the set of averages of distinct rational points in M. For a trellis $\mathscr{T} \subseteq \mathbb{N}^n$, we say that M is a $\mathscr{T}$-*rational mediated set* if $\mathscr{T} \subseteq M \subseteq \widetilde{A}(M) \cup \mathscr{T}$. Given a trellis $\mathscr{T}$ and a lattice point $\beta \in \mathrm{conv}(\mathscr{T})^\circ$, there always exists a $\mathscr{T}$-rational mediated set containing β as stated in the following Proposition.

Proposition 12.4 *Given a trellis $\mathscr{T}$ and a lattice point $\beta \in \mathrm{conv}(\mathscr{T})^\circ$, there exists a $\mathscr{T}$-rational mediated set $M_{\mathscr{T}\beta}$ containing β such that the denominators (resp. numerators) of coordinates of points in $M_{\mathscr{T}\beta}$ are odd (resp. even) numbers.*

The proof of Proposition 12.4 is constructive, and yields an algorithm to compute such a $\mathscr{T}$-rational mediated set $M_{\mathscr{T}\beta}$. We refer the reader to [MW22] for the details. By virtue of Proposition 12.4, we are able to decompose SONC polynomials into sums of binomial squares with rational exponents.

Theorem 12.5 *Let $f = \sum_{\alpha \in \mathscr{T}} c_\alpha \mathbf{x}^\alpha - d\mathbf{x}^\beta \in \mathbb{R}[\mathbf{x}]$ with $d \neq 0$ be a circuit polynomial. Assume that $M_{\mathscr{T}\beta}$ is a $\mathscr{T}$-rational mediated set containing β provided by Proposition 12.4. For each $\mathbf{u} \in M_{\mathscr{T}\beta} \setminus \mathscr{T}$, let $\mathbf{u} = \frac{1}{2}(\mathbf{v_u} + \mathbf{w_u})$ for some $\mathbf{v_u} \neq \mathbf{w_u} \in M_{\mathscr{T}\beta}$. Then f is nonnegative if and only if f can be written as $f = \sum_{\mathbf{u} \in M_{\mathscr{T}\beta} \setminus \mathscr{T}} (a_\mathbf{u} \mathbf{x}^{\frac{1}{2}\mathbf{v_u}} - b_\mathbf{u} \mathbf{x}^{\frac{1}{2}\mathbf{w_u}})^2$ for some $a_\mathbf{u}, b_\mathbf{u} \in \mathbb{R}$.*

Example 12.6 *As in Example 12.2, let $f = x_1^4 x_2^2 + x_1^2 x_2^4 + 1 - 3x_1^2 x_2^2$ be the Motzkin polynomial, $\mathscr{T} = \{\alpha_1 = (0,0), \alpha_2 = (4,2), \alpha_3 = (2,4)\}$ and $\beta = (2,2)$. Let $\beta_1 = \frac{1}{3}\alpha_1 + \frac{2}{3}\alpha_2$ and $\beta_2 = \frac{1}{3}\alpha_1 + \frac{2}{3}\alpha_3$ such that $\beta = \frac{1}{2}\beta_1 + \frac{1}{2}\beta_2$. Let $\beta_3 = \frac{2}{3}\alpha_1 + \frac{1}{3}\alpha_2$ and $\beta_4 = \frac{2}{3}\alpha_1 + \frac{1}{3}\alpha_3$. Then $M = \{\alpha_1, \alpha_2, \alpha_3, \beta, \beta_1, \beta_2, \beta_3, \beta_4\}$*

is a $\mathcal{T}$-rational mediated set containing β. By Theorem 12.5, one has

$$f = \left(a_1 x_1^{\frac{2}{3}} x_2^{\frac{4}{3}} - b_1 x_1^{\frac{4}{3}} x_2^{\frac{2}{3}}\right)^2 + \left(a_2 x_1 x_2^2 - b_2 x_1^{\frac{1}{3}} x_2^{\frac{2}{3}}\right)^2 + \left(a_3 x_1^{\frac{2}{3}} x_2^{\frac{4}{3}} - b_3\right)^2$$

$$+ \left(a_4 x_1^2 x_2 - b_4 x_1^{\frac{2}{3}} x_2^{\frac{1}{3}}\right)^2 + \left(a_5 x_1^{\frac{4}{3}} x_2^{\frac{2}{3}} - b_5\right)^2.$$

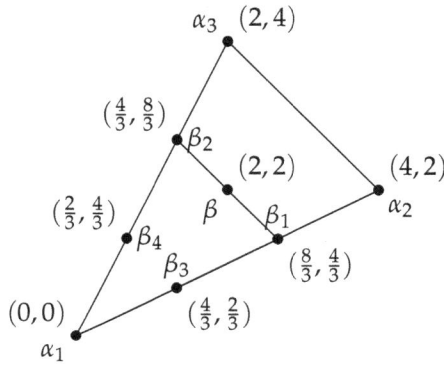

Comparing coefficients yields

$$f = \frac{3}{2}\left(x_1^{\frac{2}{3}} x_2^{\frac{4}{3}} - x_1^{\frac{4}{3}} x_2^{\frac{2}{3}}\right)^2 + \left(x_1 x_2^2 - x_1^{\frac{1}{3}} x_2^{\frac{2}{3}}\right)^2 + \frac{1}{2}\left(x_1^{\frac{2}{3}} x_2^{\frac{4}{3}} - 1\right)^2$$

$$+ \left(x_1^2 x_2 - x_1^{\frac{2}{3}} x_2^{\frac{1}{3}}\right)^2 + \frac{1}{2}\left(x_1^{\frac{4}{3}} x_2^{\frac{2}{3}} - 1\right)^2,$$

a sum of five binomial squares with rational exponents.

For a polynomial $f = \sum_{\alpha \in \mathscr{A}} f_\alpha \mathbf{x}^\alpha \in \mathbb{R}[\mathbf{x}]$, let $\Lambda(f) := \{\alpha \in \mathscr{A} \mid \alpha \in (2\mathbb{N})^n \text{ and } f_\alpha > 0\}$ and $\Gamma(f) := \text{supp}(f) \setminus \Lambda(f)$ so that $f = \sum_{\alpha \in \Lambda(f)} c_\alpha \mathbf{x}^\alpha - \sum_{\beta \in \Gamma(f)} d_\beta \mathbf{x}^\beta$. For each $\beta \in \Gamma(f)$, let

$$\mathscr{C}(\beta) := \{\mathcal{T} \mid \mathcal{T} \subseteq \Lambda(f) \text{ and } (\mathcal{T}, \beta) \text{ consists of a circuit}\}. \qquad (12.2)$$

As a consequence of Theorem 5.5 from [Wan22], if f is a SONC polynomial, then it admits a decomposition

$$f = \sum_{\beta \in \Gamma(f)} \sum_{\mathcal{T} \in \mathscr{C}(\beta)} f_{\mathcal{T}\beta} + \sum_{\alpha \in \mathscr{A}} c_\alpha \mathbf{x}^\alpha, \qquad (12.3)$$

where $f_{\mathcal{T}\beta}$ is a nonnegative circuit polynomial supported on $\mathcal{T} \cup \{\beta\}$ and $\mathscr{A} = \{\alpha \in \Lambda(f) \mid \alpha \notin \cup_{\beta \in \Gamma(f)} \cup_{\mathcal{T} \in \mathscr{C}(\beta)} \mathcal{T}\}$.

Theorem 12.7 *Let $f = \sum_{\alpha \in \Lambda(f)} c_\alpha \mathbf{x}^\alpha - \sum_{\beta \in \Gamma(f)} d_\beta \mathbf{x}^\beta \in \mathbb{R}[\mathbf{x}]$. For every $\beta \in \Gamma(f)$ and every trellis $\mathscr{T} \in \mathscr{C}(\beta)$, let $M_{\mathscr{T}\beta}$ be a $\mathscr{T}$-rational mediated set containing β provided by Proposition 12.4. Let $M = \cup_{\beta \in \Gamma(f)} \cup_{\mathscr{T} \in \mathscr{C}(\beta)} M_{\mathscr{T}\beta}$. For each $\mathbf{u} \in M \setminus \Lambda(f)$, let $\mathbf{u} = \frac{1}{2}(\mathbf{v_u} + \mathbf{w_u})$ for some $\mathbf{v_u} \neq \mathbf{w_u} \in M$. Then f is a SONC polynomial if and only if f can be written as $f = \sum_{\mathbf{u} \in M \setminus \Lambda(f)} (a_{\mathbf{u}} \mathbf{x}^{\frac{1}{2}\mathbf{v_u}} - b_{\mathbf{u}} \mathbf{x}^{\frac{1}{2}\mathbf{w_u}})^2 + \sum_{\alpha \in \mathscr{A}} c_\alpha \mathbf{x}^\alpha$ for some $a_{\mathbf{u}}, b_{\mathbf{u}} \in \mathbb{R}$.*

In order to obtain a SONC decomposition of f, we use all simplices covering β for each $\beta \in \Gamma(f)$ in Theorem 12.5. In practice, we do not need that many simplices as illustrated by the following example.

Example 12.8 *Let $f = 50x_1^4 x_2^4 + x_1^4 + 3x_2^4 + 800 - 100x_1 x_2^2 - 100x_1^2 x_2$. Let $\alpha_1 = (0,0), \alpha_2 = (4,0), \alpha_3 = (0,4), \alpha_4 = (4,4)$ and $\beta_1 = (2,1), \beta_2 = (1,2)$. There are two simplices covering β_1: the one with vertices $\{\alpha_1, \alpha_2 \alpha_3\}$ (denoted by Δ_1), and the one with vertices $\{\alpha_1, \alpha_2, \alpha_4\}$ (denoted by Δ_2). There are two simplices covering β_2: Δ_1 and the one with vertices $\{\alpha_1, \alpha_3, \alpha_4\}$ (denoted by Δ_3). One can check that f admits a SONC decomposition $f = g_1 + g_2$, where $g_1 = 20x_1^4 x_2^4 + x_1^4 + 400 - 100x_1^2 x_2$ supported on Δ_2 and $g_2 = 30x_1^4 x_2^4 + 3x_2^4 + 400 - 100x_1 x_2^2$ supported on Δ_3 are both nonnegative circuit polynomials. So the simplex Δ_1 is not needed in this SONC decomposition of f.*

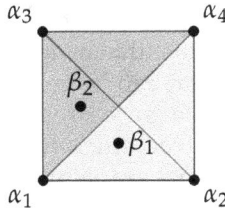

Here we rely on a heuristics to compute a set of simplices with vertices coming from $\Lambda(f)$ and that covers $\Gamma(f)$. For $\beta \in \Gamma(f)$ and $\alpha_0 \in \Lambda(f)$, let us define an auxiliary linear program:

$$\text{SimSel}(\beta, \Lambda(f), \alpha_0) := \begin{cases} \arg\max\limits_{\lambda_\alpha} & \lambda_{\alpha_0} \\ \text{s.t.} & \sum_{\alpha \in \Lambda(f)} \lambda_\alpha \cdot \alpha = \beta \\ & \sum_{\alpha \in \Lambda(f)} \lambda_\alpha = 1 \\ & \lambda_\alpha \geq 0, \quad \forall \alpha \in \Lambda(f). \end{cases}$$

If β and α_0 lie on the same face of $\mathcal{N}(f)$, the output of $\text{SimSel}(\beta, \Lambda(f), \alpha_0)$ then corresponds to a trellis which contains α_0 and covers β.

Suppose $f = \sum_{\alpha \in \Lambda(f)} c_\alpha \mathbf{x}^\alpha - \sum_{\beta \in \Gamma(f)} d_\beta \mathbf{x}^\beta \in \mathbb{R}[\mathbf{x}]$ and assume that $\{\alpha \in \Lambda(f) \mid \alpha \notin \cup_{\beta \in \Gamma(f)} \cup_{\mathscr{T} \in \mathscr{C}(\beta)} \mathscr{T}\} = \varnothing$. We first compute a simplex cover $\{(\mathscr{T}_k, \beta_k)\}_{k=1}^l$ for f by repeatedly running the program SimSel with appropriate β and α_0. Then, for each k let M_k be a $\mathscr{T}_k$-rational mediated set containing β_k and s_k be the cardinality of $M_k \setminus \mathscr{T}_k$. For each $\mathbf{u}_i^k \in M_k \setminus \mathscr{T}_k$, let us write $\mathbf{u}_i^k = \frac{1}{2}(\mathbf{v}_i^k + \mathbf{w}_i^k)$. Let $\mathbf{K}$ be the 3-dimensional rotated second-order cone, i.e.,

$$\mathbf{K} := \{(a, b, c) \in \mathbb{R}^3 \mid 2ab \geq c^2, a \geq 0, b \geq 0\}. \tag{12.4}$$

Then we can approximate $f_{\min}$ from below with the following second-order cone program:

$$\begin{cases} \sup & b \\ \text{s.t.} & f(\mathbf{x}) - b = \sum_{k=1}^l \sum_{i=1}^{s_k} (2a_i^k \mathbf{x}^{\mathbf{v}_i^k} + b_i^k \mathbf{x}^{\mathbf{w}_i^k} - 2c_i^k \mathbf{x}^{\mathbf{u}_i^k}) \\ & (a_i^k, b_i^k, c_i^k) \in \mathbf{K}, \quad \forall i, k. \end{cases} \tag{12.5}$$

12.2   First-order SDP solvers

In the previous chapters, we have explained how to reduce the size of the moment-SOS relaxations by exploiting certain sparsity structures induced by the input polynomials. A complementary framework consists of exploiting the specific properties of the matrices involved in the moment SDP relaxations, in order to speed up their resolution via specific first-order algorithms. Here, we prove that every moment relaxation of a POP with a sphere or ball constraint can be reformulated as an SDP involving a PSD matrix with constant trace property (CTP). As a result, such moment relaxations can be solved efficiently by first-order methods that exploit CTP, e.g., the conditional gradient-based augmented Lagrangian method.

First let us define CTP for a POP. Given $f, g_1, \ldots, g_m, h_1, \ldots, h_l \in \mathbb{R}[\mathbf{x}]$, let us consider the following POP with n variables, m inequality constraints and l equality constraints:

$$f_{\min} = \min \{f(\mathbf{x}) : g_j(\mathbf{x}) \geq 0, j \in [m], h_i(\mathbf{x}) = 0, i \in [l]\}. \tag{12.6}$$

Let $r_{\min} := \max \{\lceil \frac{\deg(f)}{2} \rceil, \lceil \frac{\deg(g_1)}{2} \rceil, \ldots, \lceil \frac{\deg(g_m)}{2} \rceil, \lceil \frac{\deg(h_1)}{2} \rceil, \ldots, \lceil \frac{\deg(h_l)}{2} \rceil\}$. For each $r \geq r_{\min}$, recall that the moment relaxation associated to POP (12.6) is

$$f^r = \inf_{\mathbf{y}} \left\{ L_{\mathbf{y}}(f) \; \middle| \; \begin{matrix} \mathbf{M}_r(\mathbf{y}) \succeq 0, y_0 = 1 \\ \mathbf{M}_{r-\lceil \deg(g_j)/2 \rceil}(g_j \mathbf{y}) \succeq 0, j \in [m] \\ \mathbf{M}_{r-\lceil \deg(h_i)/2 \rceil}(h_i \mathbf{y}) = 0, i \in [l] \end{matrix} \right\}. \tag{12.7}$$

If we denote

$$\mathbf{D}_r(\mathbf{y}) := \mathrm{Diag}(\mathbf{M}_r(\mathbf{y}), \mathbf{M}_{r-\lceil \deg(g_1)/2 \rceil}(g_1\mathbf{y}), \ldots, \mathbf{M}_{r-\lceil \deg(g_m)/2 \rceil}(g_m\mathbf{y})),$$

then SDP (12.7) can be rewritten as

$$f^r = \inf_{\mathbf{y}} \left\{ L_{\mathbf{y}}(f) \;\middle|\; \begin{array}{l} \mathbf{D}_r(\mathbf{y}) \succeq 0, y_0 = 1 \\ \mathbf{M}_{r-\lceil \deg(h_i)/2 \rceil}(h_i\mathbf{y}) = 0, i \in [l] \end{array} \right\}. \tag{12.8}$$

Definition 12.9 *(CTP for a POP) We say that POP (12.6) has CTP if for every $r \geq r_{\min}$, there exists $a_r > 0$ and a positive definite matrix $\mathbf{T}_r$ such that*

$$\left. \begin{array}{l} \mathbf{M}_{r-\lceil \deg(h_i)/2 \rceil}(h_i\mathbf{y}) = 0, i \in [l] \\ y_0 = 1 \end{array} \right\} \Rightarrow \mathrm{trace}(\mathbf{T}_r\mathbf{D}_r(\mathbf{y})\mathbf{T}_r) = a_r. \tag{12.9}$$

In other words, we say that a POP has CTP if each moment relaxation (12.8) has an equivalent form involving a PSD matrix whose trace is constant. In this case, we call a_r the constant trace and $\mathbf{T}_r$ the basis transformation matrix. We illustrate this conversion to an SDP with CTP.

Example 12.10 *Consider the following univariate POP*

$$-1 = \inf\{x : 1 - x^2 = 0\}.$$

Then the second-order moment relaxation is

$$f^2 = \inf_{\mathbf{y}} \; y_1$$
$$\text{s.t.} \; \begin{bmatrix} y_0 & y_1 & y_2 \\ y_1 & y_2 & y_3 \\ y_2 & y_3 & y_4 \end{bmatrix} \succeq 0, \begin{bmatrix} y_0 - y_2 & y_1 - y_3 \\ y_1 - y_3 & y_2 - y_4 \end{bmatrix} = 0, y_0 = 1.$$

It can be rewritten as

$$f^2 = \inf_{\mathbf{y}} \; y_1$$
$$\text{s.t.} \; \begin{bmatrix} 1 & y_1 & 1 \\ y_1 & 1 & y_1 \\ 1 & y_1 & 1 \end{bmatrix} \succeq 0,$$

by removing equality constraints, in which the PSD matrix has trace 3. Alternatively, with $\mathbf{D}_2(\mathbf{y}) = \mathbf{M}_2(\mathbf{y})$ and

$$\mathbf{T}_2 = \begin{bmatrix} 1 & 0 & 0 \\ 0 & \sqrt{2} & 0 \\ 0 & 0 & 1 \end{bmatrix}, \quad \mathbf{Y} = \mathbf{T}_2\mathbf{D}_2(\mathbf{y})\mathbf{T}_2 = \mathbf{T}_2\mathbf{M}_2(\mathbf{y})\mathbf{T}_2,$$

we have

$$-f^2 = \sup_{\mathbf{Y}} \{ \langle \mathbf{C}, \mathbf{Y} \rangle : \langle \mathbf{A}_i, \mathbf{Y} \rangle = b_i, i \in [5], \mathbf{Y} \succeq 0 \},$$

where $b_1 = \cdots = b_4 = 0$, $b_5 = 1$ and

$$\mathbf{C} = -\tfrac{\sqrt{2}}{4}\begin{bmatrix} 0 & 1 & 0 \\ 1 & 0 & 0 \\ 0 & 0 & 0 \end{bmatrix}, \mathbf{A}_1 = \tfrac{\sqrt{2}}{2}\begin{bmatrix} 0 & 0 & 1 \\ 0 & -1 & 0 \\ 1 & 0 & 0 \end{bmatrix}, \mathbf{A}_2 = \tfrac{1}{2}\begin{bmatrix} 2 & 0 & -1 \\ 0 & 0 & 0 \\ -1 & 0 & 0 \end{bmatrix},$$

$$\mathbf{A}_3 = \tfrac{\sqrt{2}}{4}\begin{bmatrix} 0 & 1 & 0 \\ 1 & 0 & -1 \\ 0 & -1 & 0 \end{bmatrix}, \mathbf{A}_4 = \tfrac{1}{2}\begin{bmatrix} 0 & 0 & 1 \\ 0 & 0 & 0 \\ 1 & 0 & -2 \end{bmatrix}, \mathbf{A}_5 = \begin{bmatrix} 1 & 0 & 0 \\ 0 & 0 & 0 \\ 0 & 0 & 0 \end{bmatrix}.$$

We then obtain that

$$\langle \mathbf{A}_i, \mathbf{Y} \rangle = b_i, i \in [5] \Rightarrow \text{trace}(\mathbf{Y}) = 4.$$

For the minimization of a polynomial on the unit sphere, one can show that POP (12.6) has CTP with $a_r = 2^r$ and $\mathbf{T}_r = \text{Diag}((\theta_{r,\alpha}^{1/2})_{\alpha \in \mathbb{N}_r^n})$, where $(\theta_{r,\alpha})_{\alpha \in \mathbb{N}_r^n} \subseteq \mathbb{R}^{>0}$ satisfies $(1 + \|\mathbf{x}\|_2^2)^r = \sum_{\alpha \in \mathbb{N}_r^n} \theta_{r,\alpha} \mathbf{x}^{2\alpha}$, for all $r \geq 1$.

Next, we provide a sufficient condition for POP (12.6) to have CTP. For $r \geq r_{\min}$, let $\mathcal{M}(\mathfrak{g})_r$ be the truncated quadratic module associated to $\mathfrak{g} = \{g_1, \ldots, g_m\}$. Let $\mathcal{M}^\circ(\mathfrak{g})_r$ be the interior of the truncated quadratic module $\mathcal{M}(\mathfrak{g})_r$, which is defined by

$$\mathcal{M}^\circ(\mathfrak{g})_r := \{\mathbf{v}_r^\top \mathbf{G}_0 \mathbf{v}_r + \sum_{j \in [m]} g_j \mathbf{v}_{r-\lceil \deg(g_j)/2 \rceil}^\top \mathbf{G}_j \mathbf{v}_{r-\lceil \deg(g_j)/2 \rceil}$$

$$| \mathbf{G}_j \succ 0, j \in \{0\} \cup [m]\}.$$

Theorem 12.11 *If $1 \in \mathcal{M}^\circ(\mathfrak{g})_r$ for all $r \geq r_{\min}$, then POP (12.6) has CTP. This condition holds in particular when the set of constraints includes either ball or annulus constraints.*

Once we have the knowledge of the constant a_r, the r-th order moment relaxation can be cast as follows:

$$-f^r = \sup_{\mathbf{Y}}\{\langle \mathbf{C}_r, \mathbf{Y} \rangle : \mathcal{A}_r \mathbf{Y} = \mathbf{b}_r, \mathbf{Y} \succeq 0, \text{trace}(\mathbf{Y}) = a_r\}, \qquad (12.10)$$

where $\mathcal{A}_r$ is a linear operator (used here to encode affine constraints of the SDP). Afterwards, it turns out that SDP (12.10) is equivalent to minimizing the largest eigenvalue of a matrix pencil:

$$-f^r = \inf_{\mathbf{z}}\{a_r \lambda_{\max}(\mathbf{C}_r - \mathcal{A}_r^\top \mathbf{z}) + \mathbf{b}_r^\top \mathbf{z}\}, \qquad (12.11)$$

where $\mathcal{A}_k^\top$ denotes the adjoint operator of $\mathcal{A}_k$. Hence (12.11) forms what we call a hierarchy of (nonsmooth, convex) spectral relaxations.

To solve large-scale instances of this maximal eigenvalue minimization problem, a plethora of first-order methods are available, including subgradient descent or variants of the mirror-prox algorithm, spectral bundle methods, the conditional gradient based augmented Lagrangian (CGAL) algorithm, and variants relying on limited memory and arithmetic.

12.3   Notes and sources

The material presented in Section 12.1 has been published in [WM20, MW22]. The set of SONC polynomials was introduced by [IDW16]. The condition linking nonnegativity of a circuit polynomial with its circuit number, stated in Theorem 12.1, can be found in [IDW16, Theorem 3.8]. The interested reader can find more details on mediated sets in [PR21, Rez89]. Theorem 12.3, Proposition 12.4, Theorem 12.5 correspond to Theorem 5.2 in [IDW16], Lemma 3.5 and Theorem 3.6 in [WM20], respectively. The heuristics used to compute the simplex cover can be found, e.g., in [SdW18]. Detailed numerical experiments comparing the usual geometric programming approach and the second-order cone programming approach stated in Section 12.1 can be found in [WM20, §6]. The corresponding Julia package is available at github:SONCSOCP. A recent study by [Pap19] proposes a systematic method to compute an optimal simplex cover.

The fact that SONC cones admit a second-order cone characterization was firstly stated in [Ave19, Theorem 17], but the related proof does not provide an explicit construction. Another recently introduced alternative certificates by [CS16] are sums of arithmetic geometric exponentials (SAGE), which can be obtained via relative entropy programming. The connection between SONC and SAGE polynomials has been recently studied in [MCW21, Wan22, KNT21]. It happens that SONC polynomials and SAGE polynomials are actually equivalent, and that both have cancellation-free representations in terms of generators.

The framework by [DHNdW20] relies on the dual SONC cone to compute lower bounds of polynomials by means of linear programming instead of geometric programming. Note that there are no general guarantees that the bounds obtained with this framework are always more or less accurate than the approach based on geometric programming from [SdW18], and the same holds for performance. One of the similar features shared by SOS/SONC-based frameworks is their intrinsic connections with conic programming: SOS decompositions are computed via semidefinite programming and SONC decompositions via geometric programming or second-order cone programming. In both cases, the resulting optimization problems are solved with interior-point algorithms, thus output approximate nonnegativity certificates. However, one can still obtain an exact certificate from such output via hybrid numerical-symbolic algorithms when the input polynomial lies in the interior of the SOS/SONC cone. One way is to rely on rounding-projection algorithms adapted to the SOS cone by [PP08] and the SONC cone by [MSdW19], or alternatively on perturbation-compensation schemes as in [MSEDS19, MSED18, MDSV22, MDV21, MW22].

The material on CTP presented in Section 12.2 is issued from [MML20, MLMW22]. Theorem 12.11 is proved in [MLMW22, §3]. The equivalence between SDP (12.10) and (12.11) follows from the framework by Helmberg and Rendl [HR00]. The CTP has been also studied in the context of eigenvalue optimization; see [MBM21].

CGAL is a first-order method that exploits CTP. In [YTF$^+$21], the authors combined CGAL with the Nyström sketch (named SketchyCGAL), which requires dramatically less storage than other methods and is very efficient for solving the first-order relaxation of large-scale Max-Cut instances.

When solving the second and higher order relaxations, SDP solvers often encounter the following issues:

- **Storage**: Interior-point methods are often chosen by users because of their highly accurate output. These methods are efficient for solving medium-scale SDPs. However they frequently fail due to lack of memory when solving large-scale SDPs. First-order methods (e.g., ADMM, SBM, CGAL) provide alternatives to interior-point methods to avoid the memory issue. This is due to the fact that the cost per iteration of first-order methods is much lower than that of interior-point methods. At the price of losing convexity one can also rely on heuristic methods and replace the full matrix $\mathbf{Y}$ in the moment relaxation by a simpler one, in order to save memory. For instance, the Burer-Monteiro method [BM05] considers a low rank factorization of $\mathbf{Y}$. However, to get correct results the rank cannot be too low [WW20] and therefore this limitation makes it useless for the second and higher order relaxations of POPs. Not suffering from such a limitation, CGAL not only maintains the convexity of the moment relaxation but also possibly runs with an implicit matrix $\mathbf{Y}$; see Remarks A.12 and A.17 in [MLMW22].

- **Accuracy**: Nevertheless, first-order methods have low convergence rates compared to interior-point methods. Their performance depends heavily on the problem scaling and conditioning. As a result, when solving large-scale SDPs with first-order methods it is often difficult to obtain numerical results with high accuracy.

Extensive numerical comparisons between some of these methods have been performed in §4 and Appendix A.3 from [MLMW22].

We close this chapter by emphasizing that there are other issues to be addressed to improve the scalability of polynomial optimization methods. A first complementary research track is to overcome the issue that SDP relaxations arising from the moment-SOS hierarchy can often be ill-conditioned. Possible remedies include the use of other bases of vector spaces of polynomials, for instance Chebyshev polynomials instead of the standard monomial basis as done in the univariate case in [Hen12, §5], and encoding polynomial identities by evaluating them on suitably (perhaps randomly) chosen points, seeing [LP04, LTY17] for preliminary attempts.

Another complementary research track is to exploit symmetries in polynomial optimization. For instance the invariance of all input polynomials under the action of a subgroup of the general linear group has been studied in [RTAL13]. Previous works focused on exploiting the knowledge of the group action at the SDP level [GP04]. These frameworks have been applied to compute correlation bounds for finite-dimensional quantum systems [TRR19] and bounds of packing problems [dLV15, DGV+17, DDLM21]. Exploiting symmetries has also been investigated for polynomial optimization based on sums of arithmetic-geometric-exponentials in [MNR+21]. It would be definitely worth extending these frameworks to other (non-)discrete groups, dynamical systems and to noncommutative polynomial optimization in the future. This research direction of symmetry exploitation is still to be pursued and shall hopefully lead to publishing another complementary book in the upcoming years!

Bibliography

[Ave19]   G. Averkov. Optimal size of linear matrix inequalities in
             semidefinite approaches to polynomial optimization. *SIAM
             Journal on Applied Algebra and Geometry*, 3(1):128–151, 2019.

[BM05]   Samuel Burer and Renato DC Monteiro. Local minima and
             convergence in low-rank semidefinite programming. *Math-
             ematical Programming*, 103(3):427–444, 2005.

[CS16]   V. Chandrasekaran and P. Shah. Relative Entropy Re-
             laxations for Signomial Optimization. *SIAM J. Optim.*,
             26(2):1147–1173, 2016.

[DDLM21]   Maria Dostert, David De Laat, and Philippe Moustrou. Ex-
             act semidefinite programming bounds for packing prob-
             lems. *SIAM Journal on Optimization*, 31(2):1433–1458, 2021.

[DGV$^+$17]   Maria Dostert, Cristóbal Guzmán, Frank Vallentin, et al.
             New upper bounds for the density of translative packings of
             three-dimensional convex bodies with tetrahedral symme-
             try. *Discrete & Computational Geometry*, 58(2):449–481, 2017.

[DHNdW20]   Mareike Dressler, Janin Heuer, Helen Naumann, and Timo
             de Wolff. Global optimization via the dual sonc cone and
             linear programming. In *Proceedings of the 45th International
             Symposium on Symbolic and Algebraic Computation*, pages
             138–145, 2020.

[dLV15]   David de Laat and Frank Vallentin. A semidefinite program-
             ming hierarchy for packing problems in discrete geometry.
             Mathematical Programming, 151(2):529–553, 2015.

[GP04]   Karin Gatermann and Pablo A Parrilo. Symmetry groups,
             semidefinite programs, and sums of squares. *Journal of Pure
             and Applied Algebra*, 192(1-3):95–128, 2004.

[Hen12]   Didier Henrion. Semidefinite characterisation of invariant measures for one-dimensional discrete dynamical systems. *Kybernetika*, 48(6):1089–1099, 2012.

[HR00]   Christoph Helmberg and Franz Rendl. A spectral bundle method for semidefinite programming. *SIAM Journal on Optimization*, 10(3):673–696, 2000.

[IDW16]   S. Iliman and T. De Wolff. Amoebas, nonnegative polynomials and sums of squares supported on circuits. *Research in the Mathematical Sciences*, 3(1):9, 2016.

[KNT21]   Lukas Katthän, Helen Naumann, and Thorsten Theobald. A unified framework of sage and sonc polynomials and its duality theory. *Mathematics of Computation*, 90(329):1297–1322, 2021.

[LP04]   Johan Lofberg and Pablo A Parrilo. From coefficients to samples: a new approach to sos optimization. In *2004 43rd IEEE Conference on Decision and Control (CDC)(IEEE Cat. No. 04CH37601)*, volume 3, pages 3154–3159. IEEE, 2004.

[LTY17]   JeanB Lasserre, Kim-Chuan Toh, and Shouguang Yang. A bounded degree sos hierarchy for polynomial optimization. *EURO Journal on Computational Optimization*, 5(1-2):87–117, 2017.

[MBM21]   Ngoc Hoang Anh Mai, Abhishek Bhardwaj, and Victor Magron. The constant trace property in noncommutative optimization. In *Proceedings of the 2021 on International Symposium on Symbolic and Algebraic Computation*, pages 297–304, 2021.

[MCW21]   Riley Murray, Venkat Chandrasekaran, and Adam Wierman. Newton polytopes and relative entropy optimization. *Foundations of Computational Mathematics*, pages 1–35, 2021.

[MDSV22]   Victor Magron, Mohab Safey El Din, Markus Schweighofer, and Trung Hieu Vu. Exact sohs decompositions of trigonometric univariate polynomials with gaussian coefficients. *arXiv preprint arXiv:2202.06544*, 2022.

[MDV21]   Victor Magron, Mohab Safey El Din, and Trung-Hieu Vu. Sum of squares decompositions of polynomials over their gradient ideals with rational coefficients. *arXiv preprint arXiv:2107.11825*, 2021.

[MLMW22]   Ngoc Hoang Anh Mai, Jean-Bernard Lasserre, Victor Magron, and Jie Wang. Exploiting constant trace property in large-scale polynomial optimization. *ACM Trans. Math. Softw.*, 2022.

[MML20]   Ngoc Hoang Anh Mai, Victor Magron, and Jean-Bernard Lasserre. A hierarchy of spectral relaxations for polynomial optimization. *arXiv preprint arXiv:2007.09027*, 2020.

[MNR$^+$21]   Philippe Moustrou, Helen Naumann, Cordian Riener, Thorsten Theobald, and Hugues Verdure. Symmetry reduction in am/gm-based optimization. *arXiv preprint arXiv:2102.12913*, 2021.

[MSdW19]   V. Magron, H. Seidler, and T. de Wolff. Exact optimization via sums of nonnegative circuits and arithmetic-geometric-mean-exponentials. In *Proceedings of the 2019 on International Symposium on Symbolic and Algebraic Computation*, ISSAC '19, page 291–298, New York, NY, USA, 2019.

[MSED18]   V. Magron and M. Safey El Din. On Exact Polya and Putinar's Representations. In *ISSAC'18: Proceedings of the 2018 ACM International Symposium on Symbolic and Algebraic Computation*. ACM, New York, NY, USA, 2018.

[MSEDS19]   V. Magron, M. Safey El Din, and M. Schweighofer. Algorithms for weighted sum of squares decomposition of nonnegative univariate polynomials. *Journal of Symbolic Computation*, 93:200–220, 2019.

[MW22]   Victor Magron and Jie Wang. Sonc optimization and exact nonnegativity certificates via second-order cone programming. *Journal of Symbolic Computation*, 2022.

[Pap19]   D. Papp. Duality of sum of nonnegative circuit polynomials and optimal SONC bounds. *arXiv preprint arXiv:1912.04718*, 2019.

[PP08]   H. Peyrl and P.A. Parrilo. Computing sum of squares decompositions with rational coefficients. *Theoretical Computer Science*, 409(2):269–281, 2008.

[PR21]   Victoria Powers and Bruce Reznick. A note on mediated simplices. *Journal of Pure and Applied Algebra*, 225(7):106608, 2021.

[Rez89]   B. Reznick. Forms derived from the arithmetic-geometric inequality. *Mathematische Annalen*, 283(3):431–464, 1989.

[RTAL13]   Cordian Riener, Thorsten Theobald, Lina Jansson Andrén, and Jean B Lasserre. Exploiting symmetries in sdp-relaxations for polynomial optimization. *Mathematics of Operations Research*, 38(1):122–141, 2013.

[SdW18]   H. Seidler and T. de Wolff. An experimental comparison of sonc and sos certificates for unconstrained optimization. *arXiv preprint arXiv:1808.08431*, 2018.

[TRR19]   Armin Tavakoli, Denis Rosset, and Marc-Olivier Renou. Enabling computation of correlation bounds for finite-dimensional quantum systems via symmetrization. *Physical review letters*, 122(7):070501, 2019.

[Wan22]   Jie Wang. Nonnegative polynomials and circuit polynomials. *SIAM Journal on Applied Algebra and Geometry*, 6(2):111–133, 2022.

[WM20]   Jie Wang and Victor Magron. A second order cone characterization for sums of nonnegative circuits. In *Proceedings of the 45th International Symposium on Symbolic and Algebraic Computation*, pages 450–457, 2020.

[WW20]   Irene Waldspurger and Alden Waters. Rank optimality for the burer–monteiro factorization. *SIAM Journal on Optimization*, 30(3):2577–2602, 2020.

[YTF⁺21]   Alp Yurtsever, Joel A Tropp, Olivier Fercoq, Madeleine Udell, and Volkan Cevher. Scalable semidefinite programming. *SIAM Journal on Mathematics of Data Science*, 3(1):171–200, 2021.

Part IV

Appendices: Software libraries

Appendix A

Programming with MATLAB

A.1 Sparse moment relaxations with GloptiPoly

GloptiPoly [HLL09] is a MATLAB library designed to solve the generalized problem of moments. We can rely on this library to solve the (primal) moment relaxation arising after exploiting CS for a given POP. Next we give two scripts to obtain the dense and sparse bounds given in Example 3.9. In this example, we approximate the minimum of $f = f_1 + f_2 + f_3$ on the basic compact semialgebraic set $\mathbf{X}$ with

$$
\begin{aligned}
f_1 &= -x_1 x_4, \\
f_2 &= -x_1^2 + x_1 x_2 + x_1 x_3 - x_2 x_3 + x_2 x_5, \\
f_3 &= -x_5 x_6 + x_1 x_5 + x_1 x_6 + x_3 x_6,
\end{aligned}
$$

and

$$
\mathbf{X} = \{ \mathbf{x} \in \mathbb{R}^n \mid (6.36 - x_1)(x_1 - 4) \geq 0, \ldots, (6.36 - x_6)(x_6 - 4) \geq 0 \}.
$$

By exploiting CS, the set of variables is decomposed as $I_1 = \{1, 4\}$, $I_2 = \{1, 2, 3, 5\}$, $I_3 = \{1, 3, 5, 6\}$. The first script below allows one to retrieve the dense bound f^2 of the dense primal moment relaxation (2.7) at the second relaxation order $r = 2$. Our problem involves six polynomial variables and a measure μ (defined in Line 3) supported on $\mathbf{X}$ (defined in Lines 8–13). At Line 15, GloptiPoly calls the SDP solver SeDuMi [Stu99] and returns the objective value $f^2 = 20.8608$. In addition, GloptiPoly is able to extract a minimizer of f on $\mathbf{X}$ through Algorithm 1, which then certifies that $f_{\text{min}} = f^2$.

```
1   r = 2 % relaxation order
2   mpol x1 x2 x3 x4 x5 x6
3   mu = meas([x1 x2 x3 x4 x5 x6]);
4   f1 = mom(-x1*x4);
5   f2 = mom(-x1^2 + x1*x2 + x1* x3 - x2*x3 + x2*x5);
6   f3 = mom(-x5*x6 + x1*x5 + x1*x6 + x3*x6);
7   f = f1 + f2 + f3; % objective function
8   X = [(6.36 - x1)*(x1-4)>=0, ...
9   (6.36 - x2)*(x2-4)>=0, ...
10  (6.36 - x3)*(x3-4)>=0, ...
11  (6.36 - x4)*(x4-4)>=0, ...
12  (6.36 - x5)*(x5-4)>=0, ...
13  (6.36 - x6)*(x6-4)>=0]; % support
14  Pdense = msdp(min(f), X, mass(mu)==1, r);
15  [stat,obj] = msol(Pdense);
```

For more details on the six GloptiPoly commands (mpol, meas, mom, msdp, mass, msol), we refer the interested reader to the tutorial:

$$\text{https://homepages.laas.fr/henrion/papers/gloptipoly3.pdf.}$$

The second script below allows one to retrieve the sparse bound f_{cs}^2 corresponding to the second relaxation order $r = 2$. By contrast with the dense case, one needs to define three measures μ_1, μ_2, μ_3 (Lines 6–8), associated to I_1, I_2 and I_3, respectively, and ensure that their marginals satisfy the equality constraints given in (3.5). For instance, the marginals of μ_1 and μ_2 must have equal moments for monomials in x_1 since $I_{12} = I_1 \cap I_2 = \{1\}$ (see Lines 14–15). At Line 34, GloptiPoly returns the objective value $f_{cs}^2 = f^2 = 20.8608$ and is again able to extract a minimizer of f on X.

```
1   r = 2 % relaxation order
2   mpol x1 3
3   mpol x2 x4 x6
4   mpol x3 2
5   mpol x5 2
6   mu(1) = meas([x1(1) x4]); % first measure on I1 = {1, 4}
7   mu(2) = meas([x1(2) x2 x3(1) x5(1)]); % second measure on I2 ...
        = {1, 2, 3, 5}
8   mu(3) = meas([x1(3) x3(2) x5(2) x6]); % third measure on I3 ...
        = {1, 3, 5, 6}
9   f1 = mom(-x1(1)*x4) ;
10  f2 = mom(-x1(2)^2 + x1(2)*x2(1) + x1(2)* x3(1) - x2(1)*x3(1) ...
        + x2(1)*x5(1));
11  f3 = mom(-x5(2)*x6 + x1(3)*x5(2) + x1(3)*x6 + x3(2)*x6);
12  f = f1 + f2 + f3; % objective function
13
14  m12_1 = mom(mmon(x1(1),2*r)); % moments of the marginal of ...
        mu1 on monomials in x1, corresponding to I12 = {1}
15  m12_2 = mom(mmon(x1(2),2*r)); % moments of the marginal of ...
        mu2 on monomials in x1, corresponding to I12 = {1}
16
```

```
17  m23_2 = mom(mmon([x1(2) x3(1) x5(1)],2*r)); % moments of the ...
        marginal of mu2 on monomials in x1, x3, x5, ...
        corresponding to I23 = {1, 3, 5}
18  m23_3 = mom(mmon([x1(3) x3(2) x5(2)],2*r)); % moments of the ...
        marginal of mu3 on monomials in x1, x3, x5, ...
        corresponding to I23 = {1, 3, 5}
19
20  m13_1 = mom(mmon(x1(1),2*r)); % moments of the marginal of ...
        mu1 on monomials in x1, corresponding to I13 = {1}
21  m13_3 = mom(mmon(x1(3),2*r)); % moments of the marginal of ...
        mu3 on monomials in x1, corresponding to I13 = {1}
22
23  K = [(6.36 - x1(1))*(x1(1)-4)>=0,...
24  (6.36 - x1(2))*(x1(2)-4)>=0,...
25  (6.36 - x1(3))*(x1(3)-4)>=0,...
26  (6.36 - x2)*(x2-4)>=0,...
27  (6.36 - x4)*(x4-4)>=0,...
28  (6.36 - x6)*(x6-4)>=0,...
29  (6.36 - x3(1))*(x3(1)-4)>=0,...
30  (6.36 - x3(2))*(x3(2)-4)>=0,...
31  (6.36 - x5(1))*(x5(1)-4)>=0,...
32  (6.36 - x5(2))*(x5(2)-4)>=0]; % supports
33  Psparse = msdp(min(f),m12_1-m12_2==0,m23_2-m23_3==0, ...
        m13_1-m13_3==0, K,mass(mu)==1);
34  [stat,obj] = msol(Psparse);
```

Other similar scripts for approximating minima of rational functions are given in [BHL16, §7].

A.2   Sparse SOS relaxations with Yalmip

Yalmip [Lö04] is a MATLAB toolbox to model optimization problems and to solve them using external solvers. In particular, we can rely on this toolbox to solve the SOS program (3.7), which is the dual of the moment relaxation (3.6) encoded in the prior section. Next we give two scripts to obtain the dense and sparse SOS bounds given in Example 3.9.

The first script below allows to retrieve the dense bound, corresponding to SDP (2.9) at the second relaxation order $r = 2$. For each of the six box constraints, we need to define an SOS multiplier si (see Line 11), corresponding to σ_i in (2.9). Note that the first argument si refers to the SOS polynomial multiplier itself while the second argument ci refers to its (unknown) vector of coefficients. Here, we want to maximize the lower bound b of f such that $f - b$ has a Putinar's representation of degree $2r = 4$, and thus the degree of each SOS multiplier is $2r - 2 = 2$. As with GloptiPoly, Yalmip calls at Line 15 the same SDP solver SeDuMi and returns the objective value $f^2 = 20.8608$.

```
1  r = 2; % relaxation order
```

```
2   sdpvar x1 x2 x3 x4 x5 x6 b;
3   x = [x1; x2; x3; x4; x5; x6];
4   f1 = -x1*x4;
5   f2 = -x1^2 + x1*x2 + x1* x3 - x2*x3 + x2*x5;
6   f3 = -x5*x6 + x1*x5 + x1*x6 + x3*x6;
7   f = f1 + f2 + f3; % objective function
8   g = (6.36 - x).*(x - 4); % the 6 polynomials used to define ...
        the box constraints
9   s = []; c = [b]; F = [];
10  for i=1:6
11    [si,ci] = polynomial(x,2*r-2); % SOS multiplier si of ...
          degree 2 in the dense Putinar's representation
12    s = [s si]; c = [c; ci]; F = [F, sos(si)];
13  end
14  F = [F, sos(f - b - s*g)]; % SOS multiplier s0 = f - b - ...
        sum_i si gi of degree 4
15  solvesos(F,-b,[],c); % solves the SOS program: maximize b ...
        such that f - b = s0 + sum_i si gi
```

For more details on the four Yalmip commands (`sdpvar`, `polynomial`, `sos`, `solvesos`), we refer the interested reader to the tutorial github:yalmip.

The second script below allows one to retrieve the sparse bound f^2_{cs} corresponding to the second relaxation order $r = 2$. By contrast with the dense case, one needs to define three SOS multipliers s01, s02, s03 (Lines 20–22), corresponding to $\sigma_{01}, \sigma_{02}, \sigma_{03}$ in (2.9), associated to I_1, I_2 and I_3, respectively.

As in the dense case, one needs a single SOS multiplier for each constraint but the support is more restricted. For instance, the SOS multiplier associated to the polynomial $(6.36 - x_1)(4 - x_1)$ (defined in Line 13) depends only on the variables x_1, x_4, related to the index subset $I_1 = \{1, 4\}$. At Line 28, Yalmip returns the objective value $f^2_{cs} = f^2 = 20.8608$.

```
1   r = 2; % relaxation order
2   sdpvar x1 x2 x3 x4 x5 x6 b;
3   x = [x1; x2; x3; x4; x5; x6];
4   xI1 = [x1; x4];
5   xI2 = [x1; x2; x3; x5];
6   xI3 = [x1; x3; x5; x6];
7   f1 = -x1*x4;
8   f2 = -x1^2 + x1*x2 + x1* x3 - x2*x3 + x2*x5;
9   f3 = -x5*x6 + x1*x5 + x1*x6 + x3*x6;
10  f = f1 + f2 + f3; % objective function
11  g = (6.36 - x).*(x - 4); % the 6 polynomials used to define ...
        the box constraints
12
13  [s1,c1] = polynomial(xI1,2*r-2); % SOS multiplier of degree ...
          2 in the sparse Putinar's representation, depending only ...
          on x1, x4
14  [s2,c2] = polynomial(xI2,2*r-2); % SOS multiplier of degree ...
          2 in the sparse Putinar's representation, depending only ...
          on x1, x2, x3, x5
```

```
15  [s3,c3] = polynomial(xI2,2*r-2); % SOS multiplier of degree ...
        2 in the sparse Putinar's representation, depending only ...
        on x1, x2, x3, x5
16  [s4,c4] = polynomial(xI1,2*r-2); % SOS multiplier of degree ...
        2 in the sparse Putinar's representation, depending only ...
        on x1, x4
17  [s5,c5] = polynomial(xI2,2*r-2); % SOS multiplier of degree ...
        2 in the sparse Putinar's representation, depending only ...
        on x1, x2, x3, x5
18  [s6,c6] = polynomial(xI3,2*r-2); % SOS multiplier of degree ...
        2 in the sparse Putinar's representation, depending only ...
        on x1, x3, x5, x6
19
20  [s01, c01] = polynomial(xI1,2*r); % SOS multiplier of degree ...
        4 in the sparse Putinar's representation, depending only ...
        on x1, x4
21  [s02, c02] = polynomial(xI2,2*r); % SOS multiplier of degree ...
        4 in the sparse Putinar's representation, depending only ...
        on x1, x2, x3, x5
22  [s03, c03] = polynomial(xI3,2*r); % SOS multiplier of degree ...
        4 in the sparse Putinar's representation, depending only ...
        on x1, x3, x5, x6
23
24  s = [s1 s2 s3 s4 s5 s6]; c = [b; c1; c2; c3; c4; c5; c6; ...
        c01; c02; c03];
25  F = [sos(s1), sos(s2), sos(s3), sos(s4), sos(s5), sos(s6), ...
        sos(s01), sos(s02), sos(s03)];
26
27  F = [F, coefficients(f - b - s01 - s02 - s03 - s*g,x) == 0];
28  solvesos(F,-b,[],c); % solves the SOS program: maximize b ...
        such that f - b = s01 + s02 + s03 + sum_i si gi
```

Appendix B

Programming with Julia

The goal of this chapter is to present our Julia library, called TSSOS, which aims at helping polynomial optimizers to solve large-scale problems with sparse input data. The underlying algorithmic framework is based on exploiting TS (see Chapter 7) as well as the combination of CS and TS (see Chapter 8). As emphasized in the different chapters of Part III, TS can be applied to numerous problems ranging from power networks to eigenvalue optimization of noncommutative polynomials, involving up to tens of thousands of variables and constraints. A complete documentation of the TSSOS library is available at github:TSSOS. TSSOS depends on the following Julia packages:

- MultivariatePolynomials to manipulate multivariate polynomials;

- JuMP [DHL17] to model the SDP problem;

- Graphs [FBS+21] to handle graphs;

- MetaGraphs to handle weighted graphs;

- ChordalGraph [Wan20] to generate approximately smallest chordal extensions;

- SemialgebraicSets to compute Gröbner bases.

Besides, TSSOS requires an SDP solver, which can be MOSEK [ART03] or COSMO [GCG19]. Once one of the SDP solvers has been installed, the installation of TSSOS is straightforward:

```
Pkg.add("https://github.com/wangjie212/TSSOS")
```

B.1   TSSOS for polynomial optimization

TSSOS provides an easy way to define a POP and to solve it by sparsity-adapted SDP relaxations, including the relaxation $\mathbf{P}_{ts}^{r,s}$ given in (7.17) exploiting TS, the relaxation $\mathbf{P}_{cs}^{r}$ given in (3.6) exploiting CS, as well as the relaxation $\mathbf{P}_{cs\text{-}ts}^{r,s}$ given in (8.5) exploiting both CS and TS. The tunable parameters (e.g., the relaxation order r, the sparse order s, the types of chordal extensions) allow the user to find the best compromise between the computational cost and the solution accuracy. The following Julia script is a simple example to illustrate the usage of TSSOS.

```
1 using TSSOS
2 using DynamicPolynomials
3 r = 2 /* relaxation order */
4 @polyvar x[1:6]
5 f = x[1]^4 + x[2]^4 - 2x[1]^2*x[2] - 2x[1] + 2x[2]*x[3] -
  ↪ 2x[1]^2*x[3] - 2x[2]^2*x[3] - 2x[2]^2*x[4] - 2x[2] + 2x[1]^2 +
  ↪ 2.5x[1]*x[2] - 2x[4] + 2x[1]*x[4] + 3x[2]^2 + 2x[2]*x[5] +
  ↪ 2x[3]^2 + 2x[3]*x[4] + 2x[4]^2 + x[5]^2 - 2x[5] + 2 /* objective
  ↪ function */
6 g = 1 - sum(x[1:2].^2) /* inequality constraints */
7 h = 1 - sum(x[3:5].^2) /* equality constraints */
8 nh = 1 /* number of equality constraints */
```

To solve the first step of the TSSOS hierarchy with approximately smallest chordal extensions (option TS="MD"), run

```
opt,sol,data = tssos_first([f;g;h], x, r, numeq=nh, TS="MD")
```

We obtain $f_{ts}^{2,1} = 0.2096$.
To solve higher steps of the TSSOS hierarchy, repeatedly run

```
opt,sol,data = tssos_higher!(data, TS="MD")
```

For instance, at the second step of the TSSOS hierarchy we obtain $f_{ts}^{2,2} = 0.2123$.
To solve the first step of the CS-TSSOS hierarchy, run

```
opt,sol,data = cs_tssos_first([f;g;h], x, r, numeq=nh, TS="MD")
```

to obtain the lower bound $f_{cs\text{-}ts}^{2,1} = 0.2092$.
To solve higher steps of the CS-TSSOS hierarchy, repeatedly run

```
opt,sol,data = cs_tssos_higher!(data, TS="MD")
```

For instance, at the second step of the CS-TSSOS hierarchy we obtain the lower bound $f_{\text{cs-ts}}^{2,2} = 0.2097$.
TSSOS also employs other techniques to gain more speed-up.

Binary variables

TSSOS supports binary variables. By setting nb $= a$, one can specify that the first a variables (i.e., $x_1, \ldots, x_a$) are binary variables which satisfy the equations $x_i^2 = 1$, $i \in [a]$. The specification is helpful to reduce the number of decision variables of SDP relaxations since one can identify x^j with $x^{j \,(\text{mod } 2)}$ for a binary variable x.

Equality constraints

If there are equality constraints in the description of POP (2.4), then one can reduce the number of decision variables of SDP relaxations by working in the quotient ring $\mathbb{R}[\mathbf{x}]/(h_1, \ldots, h_t)$, where $\{h_1 = 0, \ldots, h_t = 0\}$ is the set of equality constraints. To conduct the elimination, we need to compute a Gröbner basis GB of the ideal $(h_1, \ldots, h_t)$. Then any monomial $\mathbf{x}^\alpha$ can be replaced by its normal form $\text{NF}(\mathbf{x}^\alpha, GB)$ with respect to the Gröbner basis GB when constructing SDP relaxations.

Adding extra first-order moment matrices

When POP (2.4) is a QCQP, the first-order moment-SOS relaxation is also known as Shor's relaxation. In this case, $\mathbf{P}^1$, $\mathbf{P}_{\text{cs}}^1$ and $\mathbf{P}_{\text{ts}}^{1,1}$ yield the same optimum. To ensure that any higher order CS-TSSOS relaxation (i.e., $\mathbf{P}_{\text{cs-ts}}^{r,s}$ with $r > 1$) provides a tighter lower bound compared to the one given by Shor's relaxation, we may add an extra first-order moment matrix for each variable clique in $\mathbf{P}_{\text{cs-ts}}^{r,s}$. In TSSOS, this is accomplished by setting MomentOne $=$ true.

Chordal extensions

For TS, TSSOS supports two types of chordal extensions: the maximal chordal extension (option TS="block") and approximately smallest chordal extensions. TSSOS generates approximately smallest chordal extensions through two heuristics: the Minimum Degree heuristic (option TS="MD") and the Minimum Fillin heuristic (option TS="MF"). To use relaxations without TS, set TS = false. Similarly for CS there are options for the field

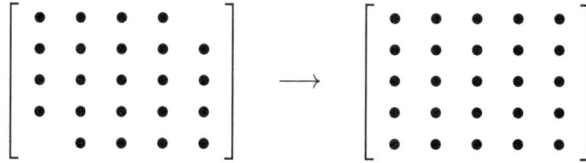

Figure B.1: Merge two 4×4 blocks into a single 5×5 block.

CS which can be "MF" by default, or "MD", or "NC" (without chordal extension), or false (without CS).

See [BK10] for a full description of the two chordal extension heuristics. The Minimum Degree heuristic is slightly faster in practice, whereas the Minimum Fillin heuristic yields on average slightly smaller clique numbers. Hence for CS, the Minimum Fillin heuristic is recommended and for TS, the Minimum Degree heuristic is recommended.

Merging PSD blocks

In case that two PSD blocks have a large portion of overlaps, it might be beneficial to merge these two blocks into a single block for efficiency. See Figure B.1 for such an example. TSSOS supports PSD block merging inspired by the strategy proposed in [GCG19]. To activate the merging process, one just needs to set the option Merge = True. The parameter md ($= 3$ by default) can be used to tune the merging strength.

Representing polynomials with supports and coefficients

The Julia package DynamicPolynomials provides an efficient way to define polynomials symbolically. But for large-scale polynomial optimization (say, $n > 500$), it is more efficient to represent polynomials by their supports and coefficients. For instance, we can represent $f = x_1^4 + x_2^4 + x_3^4 + x_1 x_2 x_3$ in terms of its support and coefficients as follows:

```
supp = [[1; 1; 1; 1], [2; 2; 2 ;2], [3; 3; 3; 3], [1; 2; 3]]
/* support array of f */
coe = [1; 1; 1; 1] /* coefficient vector of f */
```

The above representation of polynomials is natively supported by TSSOS. Hence the user can define the polynomial optimization problem directly by the support data and the coefficient data to speed up the modeling process.

SOS + sparse + RIP $\nRightarrow$ sparse SOS

To finish this section, we provide a TSSOS script showing that the relaxations based on CS can be strictly more conservative than the dense ones, even when the RIP holds. Let $f_1 = x_1^4 + (x_1 x_2 - 1)^2$, $f_2 = x_2^2 x_3^2 + (x_3^2 - 1)^2$ and $f = f_1 + f_2$. Here the RIP trivially holds as we have only two subsets of variables. After running the following commands:

```
1 using TSSOS, DynamicPolynomials
2 @polyvar x[1:3]
3 f1 = x[1]^4 + (x[1]*x[2] - 1)^2
4 f2 = x[2]^2*x[3]^2 + (x[3]^2 - 1)^2
5 f = f1 + f2
6 opt,sol,data = cs_tssos_first([f], x, 2, CS=false, TS=false)
7 opt,sol,data = cs_tssos_first([f], x, 2, TS=false)
```

we obtain $f_{cs}^2 = 0.0005 < 0.8498 = f^2$. One can also compute lower bounds based on TS as follows:

```
1 opt,sol,data = tssos_first([f], x, 2, TS="block")
2 opt,sol,data = tssos_higher!(data, TS="block")
```

to obtain $f_{ts}^{2,1} = 0.0004 < 0.8498 = f_{ts}^{2,2} = f^2$.

B.2   TSSOS for noncommutative optimization

As seen previously in Chapters 6 and 10, the whole framework of exploiting CS and TS for (commutative) polynomial optimization can be extended to handle noncommutative polynomial optimization (including eigenvalue and trace optimization), which leads to the submodule NCTSSOS in TSSOS, available at github:NCTSSOS. The corresponding commands are similar to TSSOS. To illustrate the use of NCTSSOS, we consider the eigenvalue minimization problem given in Example 6.22. After running the following commands:

```
1 @ncpolyvar x1 x2 x3 x4
2 x = [x1; x2; x3; x4]
3 f1 = 4 - x1 + 3 * x2 - 3 * x3 - 3 * x1^2 - 7 * x1 * x2 + 6 * x1 * x3
↪   - x2 * x1 -5 * x3 * x1 + 5 * x3 * x2 - 5 * x1^3 - 3 * x1^2 * x3
↪   + 4 * x1 * x2 * x1 - 6 * x1 * x2 * x3 + 7 * x1 * x3 * x1 + 2 *
↪   x1 * x3 * x2 - x1 * x3^2 - x2 * x1^2 + 3 * x2 * x1 * x2 - x2 *
↪   x1 * x3 - 2 * x2^3 - 5 * x2^2 * x3 - 4 * x2 * x3^2 - 5 * x3 *
↪   x1^2 + 7 * x3 * x1 * x2 + 6 * x3 * x2 * x1 - 4 * x3 * x2 * x2 -
↪   x3^2 * x1 - 2 * x3^2 * x2 + 7 * x3^3
```

```
4 f2 = -1 + 6 * x2 + 5 * x3 + 3 * x4 - 5 * x2^2 + 2 * x2 * x3 + 4 * x2
↪    * x4 - 4 * x3 * x2 + x3^2 - x3 * x4 + x4 * x2 - x4 * x3 + 2 *
↪    x4^2 - 7 * x2^3 + 4 * x2 * x3^2 + 5 * x2 * x3 * x4 - 7 * x2 * x4
↪    * x3 - 7 * x2 * x4^2 + x3 * x2^2 + 6 * x3 * x2 * x3 - 6 * x3 *
↪    x2 * x4 - 3 * x3^2 * x2 - 7 * x3^2 * x4 + 6 * x3 * x4 * x2  - 3
↪    * x3 * x4 * x3 - 7 * x3 * x4^2 + 3 * x4 * x2^2 - 7 * x4 * x2 *
↪    x3 - x4 * x2 * x4 - 5 * x4 * x3^2 + 7 * x4 * x3 * x4 + 6 * x4^2
↪    * x2 - 4 * x4^3
5 f = f1 + f2
6 ncball = [1 - x1^2 - x2^2 - x3^2; 1 - x2^2 - x3^2 - x4^2]
7 cs_nctssos_first([f; ncball], x, 2, CS=false, TS=false, obj="eigen")
8 cs_nctssos_first([f; ncball], x, 2, TS=false, obj="eigen")
9 cs_nctssos_first([f; ncball], x, 3, TS=false, obj="eigen")
```

we obtain $\lambda_{cs}^2(f, \mathfrak{g}) \simeq -13.7680 < -13.7333 \simeq \lambda_{cs}^3(f, \mathfrak{g}) \simeq \lambda^2(f, \mathfrak{g}) = \lambda_{\min}(f, \mathfrak{g})$.
One can also compute lower bounds based on TS as follows:

```
1 opt,data = nctssos_first([f; ncball], x, 2, TS="MD", obj="eigen")
2 opt,data = nctssos_higher!(data, TS="MD")
```

After repeating the last command twice, we obtain $\lambda_{ts}^{2,1}(f, \mathfrak{g}) \simeq -14.0922$, $\lambda_{ts}^{2,2}(f, \mathfrak{g}) \simeq -13.7618$, $\lambda_{ts}^{2,3}(f, \mathfrak{g}) \simeq -13.7393$, $\lambda_{ts}^{2,4}(f, \mathfrak{g}) = \lambda^2(f, \mathfrak{g}) = \lambda_{\min}(f, \mathfrak{g}) \simeq -13.7333$.

NCTSSOS can also handle optimization over trace polynomials. To illustrate this, let us consider the two polynomial Bell inequalities given in Section 10.4. For the first one, we use the set of noncommutative variables $(x_1, x_2, x_3, x_4) := (a_1, a_2, b_1, b_2)$, and so the objective function is

$$(\text{tr}(a_1 b_2 + a_2 b_1))^2 + (\text{tr}(a_1 b_1 - a_2 b_2))^2$$
$$= (\text{tr}(x_1 x_4 + x_2 x_3))^2 + (\text{tr}(x_1 x_3 - x_2 x_4))^2$$
$$= (\text{tr}(x_1 x_4))^2 + (\text{tr}(x_2 x_3))^2 + 2\,\text{tr}(x_1 x_4)\,\text{tr}(x_2 x_3)$$
$$+ (\text{tr}(x_1 x_3))^2 + (\text{tr}(x_2 x_4))^2 + 2\,\text{tr}(x_1 x_3)\,\text{tr}(x_2 x_4).$$

As seen earlier in the commutative setting, we represent trace polynomials by their supports and coefficients. Note that we minimize the opposite expression, so the resulting vector of coefficients is $[-1; -1; -2; -1; -1; 2]$ (see Line 4). The option constraint="unipotent" (Line 5) allows us to encode the constraints $x_i^2 = 1$ for $i \in [4]$. The next script provides an upper bound for the maximization problem given in (10.18) at relaxation order $r = 2$ and sparse order $s = 1$ with maximal chordal extensions.

```
1 n = 4
2 r = 2
```

```
3 tr_supp = [[[1;4], [1;4]], [[2;3], [2;3]], [[1;4], [2;3]], [[1;3],
  ↪  [1;3]], [[2;4], [2;4]], [[1;3], [2;4]]]
4 coe = [-1; -1; -2; -1; -1; 2]
5 opt,data = ptraceopt_first(tr_supp, coe, n, r, TS="block",
  ↪  constraint="unipotent")
```

Similarly, the next script provides an upper bound for the maximization problem (10.20) at relaxation order $r = 2$ and sparse order $s = 1$ with approximately smallest chordal extensions.

```
1 n = 6
2 r = 2
3 tr_supp = [[[1;4]], [[1], [4]], [[1;5]], [[1], [5]], [[1;6]], [[1],
  ↪  [6]], [[2;4]], [[2], [4]], [[2;5]], [[2], [5]], [[2;6]], [[2],
  ↪  [6]], [[3;4]], [[3], [4]], [[3;5]], [[3], [5]]]
4 coe = [-1; 1; -1; 1; -1; 1; -1; 1; -1; 1; 1; -1; -1; 1; 1; -1]
5 opt,data = ptraceopt_first(tr_supp, coe, n, r, TS="MD",
  ↪  constraint="unipotent")
```

To obtain bounds at higher sparse orders, one should run repeatedly

```
opt,data = ptraceopt_higher!(data, TS="block")
```

B.3   TSSOS for dynamical systems

SparseJSR is an efficient tool to compute bounds on the JSR of a set of matrices, based on the sparse SOS relaxations provided in Chapter 11. To use it in Julia, run

```
add https://github.com/wangjie212/SparseJSR
```

The following simple example illustrates how to compute upper bounds:

```
1 A = [[1 -1 0; -0.5 1 0; 1 1 0], [0.5 1 0; -1 1 0; -1 -0.5 0]]
2 r = 2 /* relaxation order */
3 ub = SparseJSR(A, r, TS = "block") /* computing an upper bound on
  ↪  the JSR of A via sparse SOS */
4 ub = JSR(A, r) /* computing an upper bound on the JSR of A via dense
  ↪  SOS */
```

As explained in Section 11.2, the JSR upper bounds are obtained via a bi-section procedure. By default, the initial lower bound and the initial upper bound for bisection are 0 and 2, respectively. The default tolerance is 1e-5. The default sparse order is 1. One can also set TS="MD" to use approximately smallest chordal extensions, which is recommended for the relaxation order $r = 1$. In addition, Gripenberg's algorithm can be employed to produce a lower bound and an upper bound on the JSR with difference at most δ.

```
1 lb,ub = gripenberg(A, δ = 0.2)
```

Finally, the SparseDynamicSystem library allows one to approximate regions of attraction, maximum positively invariant sets, global attractors for polynomial dynamic systems via the TS-based moment-SOS hierarchy. For more details, the reader is referred to the dedicated webpage: github:SparseDynamicSystem.

Bibliography

[ART03]   Erling D Andersen, Cornelis Roos, and Tamas Terlaky.   On implementing a primal-dual interior-point method for conic quadratic optimization. *Mathematical Programming*, 95(2):249–277, 2003.

[BHL16]   Florian Bugarin, Didier Henrion, and Jean Bernard Lasserre. Minimizing the sum of many rational functions. *Mathematical Programming Computation*, 8(1):83–111, 2016.

[BK10]   Hans L Bodlaender and Arie MCA Koster. Treewidth computations i. upper bounds. *Information and Computation*, 208(3):259–275, 2010.

[DHL17]   Iain Dunning, Joey Huchette, and Miles Lubin. Jump: A modeling language for mathematical optimization. *SIAM review*, 59(2):295–320, 2017.

[FBS+21]   James Fairbanks, Mathieu Besançon, Schölly Simon, Júlio Hoffiman, Nick Eubank, and Stefan Karpinski.   Juliagraphs/graphs.jl: an optimized graphs package for the julia programming language, 2021.

[GCG19]   Michael Garstka, Mark Cannon, and Paul Goulart. Cosmo: A conic operator splitting method for large convex problems. In *2019 18th European Control Conference (ECC)*, pages 1951–1956. IEEE, 2019.

[HLL09]   D. Henrion, Jean-Bernard Lasserre, and J. Löfberg. GloptiPoly 3: moments, optimization and semidefinite programming. *Optimization Methods and Software*, 24(4-5):pp. 761–779, August 2009.

[Lö04]   J. Löfberg. Yalmip : A toolbox for modeling and optimization in MATLAB. In *Proceedings of the CACSD Conference*, Taipei, Taiwan, 2004.

[Stu99]   Jos F Sturm. Using sedumi 1.02, a matlab toolbox for optimiza-
           tion over symmetric cones. *Optimization methods and software,*
           11(1-4):625–653, 1999.

[Wan20]   Jie Wang. ChordalGraph: A Julia Package to Handle Chordal
           Graphs. 2020.

www.ingramcontent.com/pod-product-compliance
Lightning Source LLC
Chambersburg PA
CBHW050600190326
41458CB00007B/2115